A-level Study Guide

Geography

Revised and updated by

Chris Burnett

David Burtenshaw

Nick Foskett

Garrett Nagle

Rosalind Foskett

Lorraine Wadsworth

Emma West

Revision Express

Acknowledgements

Garret Nagle would like to thank Angela, Rosie, Patrick and Bethany for their help.

The authors and publishers are grateful to the following for permission to reproduce copyright material:

Frontex, Europol and ICMPD for *Key migrant routes from Africa to Europe*, page 96, Guardian News and Media Ltd for *Refugees, asylum seekers, internally displaced persons (IDPs), returnees (refugees and IDPs), stateless persons and others of concern to UNHCR by country/territory of asylum, end 2007*, page 96, Unicef for *The demographic impact of educating girls*, page 99, The Financial Times for *Number of homicides by London borough, 2007*, page 127, and for *The top 20 global companies ranked by market value 2003-08*, page 138, M. Pacione, for *The dimensions of sustainability*, page 156, amended from *Sustainable Urban Development in the UK: Rhetoric or Reality*, published by the Geographical Association, Autumn 2007 and Eurostat for *Municipal waste management in the European Union*, page 168. *Ageing* and *Live births*, page 101, reproduced under the terms of the Click-Use Licence.

Every effort has been made to contact copyright holders to obtain their permission for use of copyright material. We would appreciate any information which will enable us to rectify any errors or omissions.

Series Consultants: Geoff Black and Stuart Wall

Pearson Education Limited
Edinburgh Gate, Harlow
Essex CM20 2JE, England
and Associated Companies throughout the world

© Pearson Education Limited 2001, 2008

British Library Cataloguing in Publication Data
A catalogue entry for this title is available from the British Library.

ISBN 978-1-4082-0657-7

First published 2001
Updated 2008
10 9 8 7 6 5 4

Set by Juice Creative Ltd
Printed in Malaysia (CTP-VVP)

Contents

How to use this book

Specification match

Provides a quick and easy overview of the topics that you need to study for the specification you are studying (see pages 6–7)

Exam themes

At the beginning of each chapter, these give a quick overview of the key themes that will be covered in the exam

Exam themes

- Global circulation patterns
- Air masses

The jargon

A clear outline of what subject-related and examination-related jargon means

Checkpoint

Quick question to check your understanding with full answers given at the end of the chapter

Check the net

Suggestions for useful websites related to the topic

Examiner's secrets

Hints and tips for exam success

Plate tectonics

The jargon

The *asthenosphere* is the uppermost layer of the mantle, which is semi-molten. The boundary between it and the crust is known as the *Mohorovicic* discontinuity, or *Moho*.

The earth's crust is formed out of a number of blocks or **tectonic plates** that float on a part of the mantle called the **asthenosphere**. Plate tectonics is the study of the movement of these plates and the landforms, submarine features and hazards that result from these movements.

Plate movement

You need to know the evidence that the experts used to prove that the plates move and also the mechanics of how they actually move.

Evidence for plate movement

Checkpoint 1

Can you name examples of these different pieces of evidence?

- → fit of continental coastlines
- → matching of geological structures across oceans
- → matching of climatically controlled rock formations across oceans
- → matching of species across oceans
- → palaeomagnetism
- → sea floor spreading
- → mid-ocean ridges

How plates move

Checkpoint 2

Write a short paragraph explaining how the plates of the earth's crust move.

Check the net

To read more about plate tectonics, go to www.seismo.unr.edu

Types of plate boundary

There are four types of plate boundary and the movements of the plates at each of them produce different landforms.

Constructive plate boundaries

Examiner's secrets

Your answers will be greatly improved if you know one detailed example of each of these types of plate boundary.

At this type of boundary the two plates are moving apart from each other and new crust is being created. As the plates are pulled apart by the forces of the **convection currents** in the mantle below, huge rift valleys usually form on the sea floor. Magma rises to fill any gap between the plates and a **submarine volcano** is produced. If this process continues for many years the height of the volcano may rise above sea level to create an island such as Surtsey in Iceland, which is on the Mid-Atlantic Ocean Ridge.

2

Topic checklist
A topic overview of the content covered and how it matches to the specification you are studying

Topic checklist

	Edexcel		AQA		OCR		WJEC	
	AS	A2	AS	A2	AS	A2	AS	A2
Plate tectonics	○	●		●		●	○	
Earthquakes and tsunamis	○	●		●		●	○	

Revision checklists
Allow you to monitor your progress and revise the areas that you are not confident on

By the end of this chapter you should be able to:

1	Describe the main types of pollution.	Confident	Not confident **Revise** page 82
2	Outline ways of managing pollution.	Confident	Not confident **Revise** pages 82–83

Destructive plate boundaries
There are two types of destructive boundary where two plates are moving towards each other or converging, usually causing one of them to be subducted into the asthenosphere.

1. Oceanic plate meets continental plate. The oceanic plate subducts under the continental plate and a deep ocean trench forms. Sediments in the area are folded and faulted into a range of fold mountains along the edge of the continental plate. Some 80% of the world's volcanoes are located in these areas.
2. Two oceanic plates converge. When two oceanic plates converge the result is a deep ocean trench and a line of volcanic islands called an island arc.

Collision plate boundaries
Where two continental plates converge, which is unusual, the two plates are forced together and as they are of similar density and much lighter than the asthenosphere, no subduction takes place. The collision forces the sediments between the two plates to be forced upwards into some of the largest mountain ranges in the world.

Conservative plate boundaries
At this type of boundary the plates are being dragged past each other and no crustal material is being created or destroyed. Where this type of movement is taking place there is often a series of **transform faults**. These are at right angles to the main boundary and earthquakes are a major hazard in these locations as much as along the main boundary.

The crust
There are two types of crust making up the outer layer of the earth.

→ Oceanic crust (sima) is almost continuous around the earth, is relatively dense (3.0–3.3 gm per cc), consists mainly of basaltic rocks and is less than 250 million years old. This forms only a thin layer between 6 and 10 km thick.
→ Continental crust (sial) only occurs where there are continental land masses (on average it is between 35 and 70 km thick), is lighter, only about 2.7 gm per cc, consists mainly of granitic rocks and is much older – in some places more than 3500 million years old. The oldest, most stable and most eroded areas of this type of crust are called **cratons or shield areas**.

Action point
Draw and label a diagram of each type of destructive plate boundary to explain the processes taking place.

Test yourself
Can you accurately locate an example of each of these types of plate boundary?

Links
See tectonic hazards, pages 72–73.

Watch out!
Often both plates are moving in the same direction at one of these plate boundaries, but at a different speed. This gives the impression that they are actually moving in opposite directions.

Grade booster
The secret of reaching the highest grade in questions about these topics, can be to demonstrate that you understand that the moving plates are all part of a continuous process, and not just isolated, one-off events.

Exam practice answers: page 16
1. Using the theories of plate tectonics, explain the major relief features of one continent that you have studied. (20 mins)
2. There are many pieces of evidence that scientists have used to prove the theories of plate tectonics. Briefly describe the most important of these. (10 mins)

3

Action point
A suggested activity linked to the content

Test yourself
Quick memory recall activities

Links
Cross-reference links to other relevant sections in the book

Watch out!
Flags up common mistakes and gives hints on how to avoid them

Grade booster
Examiner suggestions on how to get the top grade

Exam practice
Exam-style questions to check your understanding of the topic content with full answers given at the end of the chapter

Specification map

Plate tectonics	Plate tectonics
	Earthquakes and tsunamis
	Volcanicity
Climate	Weather and climate
	Climate change
	Issues of climate change
	Coping with climate change
	Microclimates
Water	Hydrological processes
	Rivers and their landforms
Biomes and ecosystems	Tropical forest and grassland ecosystems
	Ecosystem processes and structures
	Small-scale ecosystems
	Ecological succession
	Threats to biodiversity
Environments	Coastal environments
	Threats to coasts
	Tundra and polar environments
	Glacial environments – glacial systems
	Glacial environments – glacial landforms
	Desert environments
Natural hazards	Principles and slope hazards
	Tectonic hazards
	Climatic hazards
	Flood hazards and water conflicts
Pollution	Principles of pollution
	Water pollution
	Atmospheric pollution: acid rain
Population	Demography and population
	Migration
	Refugees and asylum seekers
	Gender issues
	Demographic challenges
Settlement	Urbanisation
	Regeneration and reurbanisation
	The central business district
	Social structures and diversity
	Inner-city issues
	Retailing and services
	The urban fringe
	Rural settlement changes
	Leisure and recreation
	Deprivation
	Geography of crime
	Perceiving environments
Economic change	Globalisation
	Transnational corporations
	Levels of development
	Inequalities
	The development gap
Sustainability	Principles of sustainability
	Sustainable environments
	Sustainable food supply
	Sustainable cities
	Sustainable water supplies
	Energy issues
	Sustainable energy supplies
	Sustainable tourism
	Health and welfare
	Waste disposal
Global geography	Superpower geography: China
	Superpower geography: India
	Newly-industrialising countries (NICs)

Edexcel		AQA		OCR		WJEC	
AS	A2	AS	A2	AS	A2	AS	A2
○	●		●		●	○	
○	●		●		●	○	
○	●		●		●	○	
○					●	○	
○					●	○	●
○					●	○	
○					●	○	
○					●	○	
		○		○		○	
		○		○		○	
			●				
	●		●		●		●
			●		●		●
	●		●		●		●
	●		●		●		●
○		○		○			●
○		○		○			●
	●	○		○			●
	●	○		○			●
	●	○		○			●
	●	○		○			●
			●		●	○	
○	●		●		●	○	
○			●		●		●
○			●		●	○	
	●		●		●		●
	●		●		●		●
	●		●		●		●
○		○			●	○	
○		○			●	○	
○					●	○	
	●					○	
○		○			●	○	
○	●	○		○		○	
	●	○	●	○	●	○	●
○	●		●	○		○	●
○	●	○	●	○	●	○	●
○	●	○			●	○	●
	●	○			●	○	●
○		○		○		○	
○						○	
		○	●			○	●
			●			○	●
	●	○	●	○		○	●
							●
	●		●				●
	●		●				●
	●		●				●
	●		●				●
	●		●				●
○		○		○			●
				○			●
		○					●
				○			●
							●
		○			●		●
		○			●		●
							●
		○					●
○							●
○	●			○	●		●
○	●			○	●		●
○	●	○	●	○	●		●

Plate tectonics

The study of Geography requires you to learn about the structure of the earth and plate tectonics. Having first looked at the structure of the earth's crust and the way that it is changing, you will study the landforms that are created by the enormous forces of nature which are released at the plate boundaries. You will develop an understanding of the four different types of plate boundary and their location. Earthquakes, volcanoes and tsunamis are found at these locations and you will study their causes and characteristics, The approach adopted is to investigate the landforms and processes that are responsible for their creation. With all questions it is vital that you can explain how the processes that have caused the landform operate, as well as to know real-world examples of places where these things have actually taken place.

Exam themes

- Plate movement
- Types of plate boundary
- Characteristics of the crust
- Earthquakes
- Tsunamis
- Intrusive and extrusive landforms including types of volcano

Topic checklist

	Edexcel		AQA		OCR		WJEC	
	AS	A2	AS	A2	AS	A2	AS	A2
Plate tectonics	○	●		●		●	○	
Earthquakes and tsunamis	○	●		●		●	○	
Volcanicity	○	●		●		●	○	

Plate tectonics

The jargon

The *asthenosphere* is the uppermost layer of the mantle, which is semi-molten. The boundary between it and the crust is known as the *Mohorovicic* discontinuity, or *Moho*.

The earth's crust is formed out of a number of blocks or **tectonic plates** that float on a part of the mantle called the **asthenosphere**. Plate tectonics is the study of the movement of these plates and the landforms, submarine features and hazards that result from these movements.

Plate movement

You need to know the evidence that the experts used to prove that the plates move and also the mechanics of how they actually move.

Evidence for plate movement

→ fit of continental coastlines
→ matching of geological structures across oceans
→ matching of climatically controlled rock formations across oceans
→ matching of species across oceans
→ **palaeomagnetism**
→ **sea floor spreading**
→ **mid-ocean ridges**

Checkpoint 1

Can you name examples of these different pieces of evidence?

How plates move

Checkpoint 2

Write a short paragraph explaining how the plates of the earth's crust move.

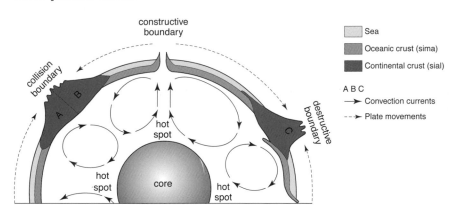

Types of plate boundary

There are four types of plate boundary and the movements of the plates at each of them produce different landforms.

Constructive plate boundaries

At this type of boundary the two plates are moving apart from each other and new crust is being created. As the plates are pulled apart by the forces of the **convection currents** in the mantle below, huge rift valleys usually form on the sea floor. Magma rises to fill any gap between the plates and a **submarine volcano** is produced. If this process continues for many years the height of the volcano may rise above sea level to create an island such as Surtsey in Iceland, which is on the Mid-Atlantic Ocean Ridge.

Examiner's secrets

Your answers will be greatly improved if you know one detailed example of each of these types of plate boundary.

Destructive plate boundaries

There are two types of destructive boundary where two plates are moving towards each other or converging, usually causing one of them to be subducted into the asthenosphere.

1   Oceanic plate meets continental plate. The oceanic plate subducts under the continental plate and a deep ocean trench forms. Sediments in the area are folded and faulted into a range of fold mountains along the edge of the continental plate. Some 80% of the world's volcanoes are located in these areas.
2   Two oceanic plates converge. When two oceanic plates converge the result is a deep ocean trench and a line of volcanic islands called an island arc.

Collision plate boundaries

Where two continental plates converge, which is unusual, the two plates are forced together and as they are of similar density and much lighter than the asthenosphere, no subduction takes place. The collision forces the sediments between the two plates to be forced upwards into some of the largest mountain ranges in the world.

Conservative plate boundaries

At this type of boundary the plates are being dragged past each other and no crustal material is being created or destroyed. Where this type of movement is taking place there is often a series of **transform faults**. These are at right angles to the main boundary and earthquakes are a major hazard in these locations as much as along the main boundary.

The crust

There are two types of crust making up the outer layer of the earth.

➔   Oceanic crust (sima) is almost continuous around the earth, is relatively dense (3.0–3.3 gm per cc), consists mainly of basaltic rocks and is less than 250 million years old. This forms only a thin layer between 6 and 10 km thick.
➔   Continental crust (sial) only occurs where there are continental land masses (on average it is between 35 and 70 km thick), is lighter, only about 2.7 gm per cc, consists mainly of granitic rocks and is much older – in some places more than 3500 million years old. The oldest, most stable and most eroded areas of this type of crust are called **cratons or shield areas**.

Action point

Draw and label a diagram of each type of destructive plate boundary to explain the processes taking place.

Checkpoint 3

Can you accurately locate an example of each of these types of plate boundary?

Watch out!

Often both plates are moving in the same direction at one of these plate boundaries, but at a different speed. This gives the impression that they are actually moving in opposite directions.

Grade booster

The secret of reaching the highest grade in questions about these topics, can be to demonstrate that you understand that the moving plates are all part of a continuous process, and not just isolated, one-off events.

Exam practice   answers: page 16

1   Using the theories of plate tectonics, explain the major relief features of one continent that you have studied. (20 mins)
2   There are many pieces of evidence that scientists have used to prove the theories of plate tectonics. Briefly describe the most important of these. (10 mins)

Earthquakes and tsunamis

Earthquakes and tsunamis can be two of the most devastating natural events. The vibrations caused by earthquakes can be a mere inconvenience in MEDCs who have the technology and support systems to mitigate their effects, whereas they are frequently devastating in their effects in LEDCs, where the populations are less able to cope with their effects. Tsunamis can be caused by earthquakes and can also have devastating effects.

Earthquakes

Earthquakes happen when stresses in the earth's crust reach such a level that rocks either break apart or suddenly move past each other. The point where the rocks break or fracture is referred to as the **focus**, which is usually on a fault line. The **epicentre**, the point on the earth's surface directly above the focus, usually experiences most vibration.

Energy is released from the focus in a series of different types of waves, the study of which has enabled scientists to develop a greater understanding of the internal structure of the earth. Most earthquakes occur at plate boundaries, the majority of the largest take place at subduction and collision zones but they also happen at mid-ocean ridges and along conservative plate boundaries such as the San Andreas Fault. As many earthquakes occur under the sea, **tsunamis** are an associated hazard in coastal areas. Minor earthquakes also occur:

→ near volcanic eruptions
→ along old fault lines, often due to isostatic readjustment
→ when stresses in the crust have been released by filling valleys with very heavy loads of water when building huge reservoirs.

Measuring earthquakes

There are two scales used to measure earthquakes. The **Richter scale** is a logarithmic scale, which goes up to 10, of the amount of energy released. The **Mercalli scale** is a measure of the amount of ground movement that occurs.

Tsunamis

Tsunamis are huge ocean waves that are usually caused by plate movements on the ocean floor, but can also be caused by volcanic eruptions or large landslides. They are actually a series of waves that travel outwards from the source of the movement like the ripples on the surface of a pond when a stone is dropped in. They may occur in any large area of water, but most are in the Pacific Ocean, the name itself is a Japanese word meaning "Harbour Wave".

Tsunamis can be devastating because:

→ They travel at high speed, up to 950km/hour over open seas and are difficult to detect from the air, as they are only a few metres high.
→ Their height increases and their speed decreases as the water over which they are travelling gets shallower.
→ They hit the shore as a wall of water up to 50m high, travelling at about 50km/hour.
→ The waves are so powerful that they can uproot trees and carry buildings, boulders and cars along with them.
→ The damaging effects are felt some considerable distance inland, particularly where the land is low lying and flat. Areas most at risk are within 1 to 2 kms inland and below 10 metres above sea level.

The approach of a tsunami is usually heralded by an unusual series of alternating high and low water levels several minutes apart. There may also be some ground movement if the earthquake is near the shoreline. As the waves refract around islands, all sides are in danger of their effects, not just the side facing the source of the tsunami.

Action point

Research the causes and effects of the devastating tsunami of Boxing Day 2004.

Check the net

There is a website dedicated to the Asian Tsunami of 2004 – check it out at www.tsunami2004.net

Exam practice   answers: page 16

1 Why do earthquakes associated with the Mid-Atlantic Ridge usually have a shallow focus whereas those in Japan have a deep focus? (10 mins)
2 Explain the three ways in which tsunamis can be formed. (10 mins)

Volcanicity

Volcanicity refers to the forcing of solid, liquid or gaseous material into or onto the surface of the earth's crust. These processes are referred to as **intrusive** when the materials are forced into the crust and **extrusive** when the material is forced onto the surface.

Intrusive landforms

Far more magma is intruded into the crust than ever reaches the surface. It cools and solidifies within the crust and after many centuries of **denudation** the various forms of intrusion are exposed as landforms. These vary in size and shape, often depending on the rock structures into which the intrusions take place, and include **batholiths**, **laccoliths**, **phacoliths**, **dykes** and **sills**. As the hot material is injected into the crust from below, the existing solid rock experiences changes due to the heat and/or pressure exerted upon it. The rocks that are changed in this way are referred to as the **metamorphic aureole**.

These landforms are usually visible only when centuries of denudation have taken place and overlying layers of rock have been removed. Intrusive volcanic events are only hazardous through the earthquakes which accompany such intrusions.

Extrusive landforms

Major landforms

Volcanoes are the most common landform that result from extrusive volcanic activity and they may be classified in a number of ways:

→ the characteristics of the extruded material (acidic or basic)
→ the nature of the opening or vent through which the lava has emerged (fissure or vent)
→ the frequency of the eruption at that location (regular or infrequent)
→ the degree of violence of the eruption (explosive or gentle).

Volcanoes are also classified according to the name of a well known volcano which has erupted in a similar way. Examples of these are **Strombolian**, **Vulcanian**, **Vesuvian**, **Pelean**, and **Hawaiian**. The eruptions which are often the most powerful are referred to as **Plinian**. This type of eruption is explosive and extrudes viscous lava. An example of this type of eruption was Mount St Helens in 1980.

Volcanoes may occur in a variety of different shapes and sizes depending upon the factors above.

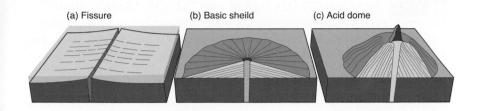

(a) Fissure   (b) Basic sheild   (c) Acid dome

(d) Ash cinder   (e) Composite   (f) Caldera

Take note

Make copies of each of the different volcanoes in the diagram and add as many labels as you can to explain their characteristic shapes.

Where are volcanoes found?

Volcanoes are found in five different types of location:

Checkpoint 1

Explain why volcanoes are found at each of the locations listed.

→ at spreading mid-ocean ridges
→ along continental rift valleys
→ in island arcs at destructive plate margins
→ in fold mountain ranges
→ in isolated locations known as hot spots.

Minor landforms

Minor landforms such as **mud volcanoes**, **solfataras**, **geysers** and **fumaroles** are formed as a result of extrusive volcanic activity.

Test yourself

You should be able to describe and explain the formation of each of these minor extrusive features.

Other events associated with volcanic eruptions

The eruption of a volcano can result in other, often hazardous, events depending on where the volcano is situated. In Iceland, where many glaciers overlay volcanoes, the word **jökulhlaup** means a sudden outburst of meltwater caused by the ice melting as a result of volcanic activity. These events can cause devastating sudden floods leading to infrastructure damage and loss of life.

Action point

Research an example of a jökulhlaup.

In 1984 the town of Armero in Colombia was destroyed by a series of lahars, which emanated from the icefields on top of the Nevada del Ruiz volcano. A lahar is a type of mudflow containing pyroclastic material from a volcano mixed with water, usually meltwater from snow and ice or from local lakes. The Mount St Helens eruption of 1980 also led to a series of lahars.

Checkpoint 2

Explain how lahars are formed.

Exam practice   answers: page 16

1 Discuss the opinion that volcanicity can be said to be both hazardous and advantageous to people. (20 mins)
2 Most volcanoes are located at, or close to, plate boundaries, but the islands of Hawaii are volcanoes in the centre of the Pacific plate. Explain this apparent anomaly. (10 mins)

Examiner's secrets

As with all aspects of geography it is vital for you to know a detailed example of each feature or event you have learnt about.

Answers
Plate tectonics

Checkpoints

1 Fit of coastlines – Africa and South America; matching geological structures in SE Brazil and S Africa; climatically controlled rock formations – coal in Antarctica; small Permian reptiles found only in Africa and Brazil; palaeomagnetism, sea floor spreading, mid-ocean ridges all along the Mid-Atlantic Ridge.

2 Convection currents in the mantle, which are generated by heat from the centre of the earth, move the solid plates that float on this material. You should add a section about the way that the plates move at each of the three types of plate boundary.

3 Constructive: Two plates are diverging at the Mid-Atlantic ridge. Destructive: Oceanic plate meets continental off west coast of S America where Nazca goes under South American plate. Destructive: Two oceanic plates converge in Western Pacific. Collision: Two continental plates converge in India.

Exam practice

1 North America or Europe are most likely. You should be able to describe young and old fold mountain chains, cratons, geosyncline, and possibly a rift valley.

2 You should describe the way that at least three of the pieces of evidence for plate movement lead people to think that the theory of plate tectonics is the only possible explanation for these phenomena. It is vital to include located examples in your answer.

Earthquakes and tsunamis

Checkpoint

Compressional, shear (both high frequency), and Love and Rayleigh (low frequency).

Exam practice

1 In Japan, earthquakes have a deep focus because the subduction zone, where the foci are, is beneath the thick continental plate. Earthquakes associated with the Mid-Atlantic Ridge usually have a shallow focus as the movements are only associated with the shallow ocean plates.

2 Tsunamis are huge ocean waves which are usually caused by plate movements on the ocean floor, but can also be caused by volcanic eruptions or large landslides. They are actually a series of waves which travel outwards from the source of the movement like the ripples on the surface of a pond when a stone is dropped in. They travel at high speed, up to 950km/hour and are difficult to detect from the air, as they are only a few metres high. They may occur in any large area of water, but most are in the Pacific Ocean, the name itself is a Japanese word meaning "Harbour Wave".

Volcanicity

Checkpoints

1 At spreading mid-ocean ridges as magma rises between spreading plates. Along continental rift valleys the brittle crust is fractured allowing magma to rise from below. In island arcs and in fold mountain ranges magma rises under pressure created from the melting lithosphere. In isolated locations known as hot spots, volcanoes are due to a rising plume of molten material.

2 Lahars are mudflows or water-saturated pyroclastic debris on volcanoes. They form when the pyroclastic flows associated with a volcanic eruption cause snow or glaciers to melt or when an eruption causes a lake to overflow and mix with other deposits.

Exam practice

1 Hazardous consequences are to do with the disastrous effects. They include loss of life, damage to property, stress, loss of employment, etc. Advantageous effects include gaining a better knowledge of mineral sources, the use of rich soils from volcanic ash and lava breakdown, creation of new land, geothermal energy, minerals and tourism.

2 The volcanoes of Hawaii are hot spot volcanoes and are the only type not found near a plate boundary. In your answer you need to be able to explain that a hot spot is a plume of rising magma beneath the crust. The magma rises through the mantle, and as it is lighter than the surrounding rock it erupts onto the ocean floor. After many eruptions, sufficient layers have been deposited to allow the lava to rise above sea level. In the case of Hawaii this has been in the same location for millions of years and the Pacific Plate has moved over it causing a series of volcanoes, which we know as the Hawaiian islands.

Revision checklist
Plate tectonics

By the end of this chapter you should be able to:

1	Describe and explain how tectonic plates move.	Confident	Not confident **Revise** pages 10–11
2	Describe and explain the landforms associated with each of the four types of plate boundary.	Confident	Not confident **Revise** pages 10–11
3	Describe the main characteristics of the earth's crust.	Confident	Not confident **Revise** pages 10–11
4	Describe and explain what causes earthquakes.	Confident	Not confident **Revise** pages 12–13
5	Describe and explain what causes tsunamis.	Confident	Not confident **Revise** pages 12–13
6	Describe and explain the formation of intrusive and extrusive igneous landforms.	Confident	Not confident **Revise** pages 14–15
7	Describe and explain some other events that are associated with volcanic eruptions.	Confident	Not confident **Revise** pages 14–15

Climate

Weather and climate are vital to the study of physical geography and human geography. Weather and climate are important in their own right but also influence society through agriculture, tourism, hazards, insurance claims, water supplies and so on. Understanding weather and climate is essential for understanding how people manage the environment in which they live. Moreover, in an increasingly urban–industrial world, the experience of weather and climate is most people's only contact with the natural environment.

Exam themes

- Global circulation patterns
- Air masses
- Low pressure systems
- High pressure systems
- Weather hazards
- Contrasting climate (monsoon, tropical and temperate)
- Small-scale climate, e.g. microclimate

Topic checklist

	Edexcel		AQA		OCR		WJEC	
	AS	A2	AS	A2	AS	A2	AS	A2
Weather and climate	○			●		●	○	
Climate change	○			●		●	○	●
Issues of climate change	○			●		●	○	
Coping with climate change	○			●		●	○	
Microclimates	○			●		●	○	

Weather and climate

Check the net

Look at www.bbc.co.uk/weather/
features/understanding/monsoon.shtml
for a good site about monsoons.

The term **climate** refers to the state of the atmosphere over a period of not less than thirty years. It includes variables such as temperature, rainfall, winds, humidity, cloud cover, and pressure. It refers not just to the averages of these variables but to the extremes as well. By contrast, **weather** refers to the state of the atmosphere at any particular moment in time. However, we usually look at the weather over a period of between a few days and a week. The same variables as for climate are considered.

Climate associated with hot deserts

Climate data for Cairo

	J	F	M	A	M	J	J	A	S	O	N	D	Year
Average monthly temperature (°C)	14	15	18	21	25	28	29	28	26	24	20	16	22
Monthly total rainfall (mm)	4	4	3	1	2	1	0	0	1	1	3	7	27

The main characteristics of hot desert climates include:

→ very hot days and cold nights, caused by the lack of cloud cover
→ low and irregular amounts of rainfall, which lack any seasonal pattern
→ low levels of humidity for much of the year
→ warm dry winds, sometimes causing sandstorms, can be a feature of hot desert climates.

Monsoon climates

Checkpoint 1

What does the term *monsoon* mean?

Checkpoint 2

What is a *land–sea breeze*?

The **monsoon** is the reversal of pressure and winds that gives rise to a marked seasonality of rainfall over north and south-east Asia. This can be seen clearly in India and Bangladesh. The monsoon has sometimes been described as a giant **land–sea breeze** – but this is too much of a simplification. A number of influences have been suggested:

→ differential heating and cooling of land compared with the adjacent seas
→ seasonal movement of the **inter-tropical convergence zone** (ITCZ)
→ the effect of the Himalayas on the ITCZ

Winter monsoon

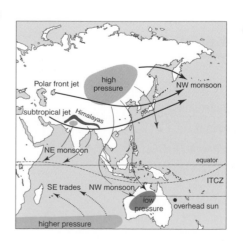

→ Temperatures over central Asia are low, leading to high pressure.
→ The jet stream splits into two: the southern subtropical jet leads to descending air and high pressure.
→ This brings outward-blowing north-easterly winds across south Asia.
→ An intense low pressure area develops over northern Australia, where it is summer and very warm.
→ Winds blow from the Asian high pressure area to the more intense Australian low pressure area.

Checkpoint 3

What is a *jet stream*?

Summer monsoon

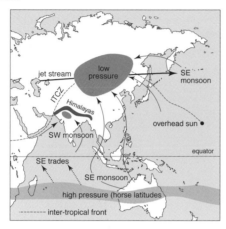

→ In March and May the winds shift, and the upper westerly air currents begin to move north.
→ The **jet stream** strengthens until it lies entirely to the north of the Himalayas.
→ The overhead sun migrates north to a position just over India, and the **ITCZ** moves north (the monsoon trough).
→ Intense low pressure develops over Asia. This is separated, by the Himalayas, from a smaller area of intense low pressure over the Punjab.
→ High pressure develops over northern Australia, where it is winter; winds blow from the Australian high pressure area to the more intense Asian low pressure area.
→ Climate data for Cherrapunji is given below.

Action point

Plot the climate data shown in the two tables and describe the differences in the climates at Cherrapunji and Cairo.

Grade booster

Quote figures. Use the data in the table to show how much rain falls in the rainy season, and how much in the dry season. For top marks, convert this to a percentage of the year's total.

	J	**F**	**M**	**A**	**M**	**J**	**J**	**A**	**S**	**O**	**N**	**D**
Temp. (°C)	11	13	15	17	18	19	20	21	20	19	16	11
Rain (mm)	24	38	170	595	1700	290	270	1700	1200	420	40	10

Exam practice   answer: page 30

Describe and explain the weather patterns associated with the Indian monsoon. (30 mins)

Climate change

The jargon

The lower atmosphere acts like a *greenhouse*, allowing short-wave radiation in, but trapping outgoing long-wave radiation.

The 'greenhouse effect' is the process by which certain gases absorb outgoing long-wave radiation from the earth and return some of it back to earth. In all, greenhouse gases raise the earth's temperatures by about 33°C.

Properties of greenhouse gases

Greenhouse gases vary in their abundance and contribution to global warming.

Check the net

View the Friends of the Earth climate change website at:
www.foe.co.uk/campaigns/climate/issues/climate_change_index.html

Greenhouse gas	Average atmospheric concentration (ppm)	Rate of change (% per annum)	Direct global warming potential (GWP)	Lifetime (years)	Type of indirect effect (ppm)
Carbon dioxide	355	0.5	1	120	None
Methane	1.72	0.6–0.75	11	10.5	Positive
Nitrous oxide	0.31	0.2–0.3	270	132	Uncertain
CFC 11	0.000255	4	3400	55	Negative
CFC 12	0.000453	4	7100	116 months	Negative
CO					Positive
NO$_x$					Uncertain

The rise of greenhouse gases and global warming

Carbon dioxide levels rose from about 315 ppm in 1950 to 355 ppm in 1999 and are expected to reach 600 ppm by 2050. The increase is due to human activities:

→ burning **fossil fuels** – coal, oil and natural gas
→ **deforestation** of the tropical rainforest is a double blow – not only does it increase atmospheric CO_2 levels, but it also removes the trees that convert CO_2 into oxygen
→ **methane** is the second largest contributor to global warming – cattle convert up to 10% of the food they eat into methane, and emit 100 million tonnes of methane into the atmosphere each year

Checkpoint 1

Why is the build-up of methane and CFCs of greater concern than the build-up of carbon dioxide?

→ **chlorofluorocarbons** (CFCs) are synthetic chemicals that destroy ozone and absorb long-wave radiation – CFCs are increasing at a rate of 6% per annum, and are up to 10 000 times more efficient at trapping heat than CO_2

Evidence for the greenhouse effect is the link between CO_2 levels and temperatures (modern records only began in 1860):

→ average temperatures have risen by over 0.5°C over the last century
→ the last decade is the warmest on record

→ nine of the ten warmest years in Western Europe have occurred since 1983.

The effects of global warming

→ Sea levels will rise, causing flooding in low-lying areas such as the Netherlands, Egypt and Bangladesh.
→ There will be an increase in storm activity (because there will be more atmospheric energy).
→ Agricultural practices and patterns will change.
→ There will be a different distribution of rainfall.

Ways of reducing greenhouse gas emissions

The energy sector

→ Introduce a carbon tax on electricity generation.
→ Introduce a higher CO_2 tax and retain the existing energy tax on non-energy-intensive industry.

The transport sector

→ New rules and taxes on company cars (to reduce long-distance travel).
→ Expand public transport systems.
→ Set CO_2 emission limits on light vehicles.

Other greenhouse gases

→ Reduce agricultural use of nitrogen fertilisers.
→ Expand methane extraction from waste tips.

The ozone hole

Ozone (O_3) is continuously created and destroyed in the atmosphere. Oxygen (O_2) is broken down into individual atoms by ultraviolet radiation. Some of these atoms combine with oxygen to form ozone. Ozone is broken down by ultraviolet radiation – so there is a natural cycle of growth and decay.

Ozone is important as it filters out harmful **ultraviolet (UV) radiation**. However, when CFCs are broken down in the atmosphere they release chlorine, which destroys ozone. As this destruction of ozone is faster than its natural regeneration, the amount of ozone is decreasing.

The effects of reduced O_3 include:

→ increased risk of skin cancer
→ more eye diseases such as cataracts
→ crop yields will decline by 25% (if O_3 declines by 25%).

The ozone 'hole' is a large area over Antarctica (and to a lesser extent over the Arctic) where there are less than half normal levels of O_3.

Checkpoint 2

How is ozone broken down and re-formed?

Grade booster

Don't be too general. Refer to figures provided by research agencies or the government to explain what might happen as a result of the greenhouse effect and global warming. Appreciate that pressure groups might use the 'worst case scenario' to make their point.

Exam practice   answers: page 30

1   What is the 'greenhouse effect'? (10 mins)
2   How could global warming increase storm frequency and intensity? (10 mins)

Issues of climate change

There are many issues related to climate change – environmental, economic, social and political, depending on how great the changes are. These can also be further sub-divided into short-term changes, long-term changes, direct impacts and indirect impacts. Moreover, another issue is whether climate change is due to global warming or global dimming.

Environmental issues

These include:

→ increased flood frequency and flood magnitude
→ more frequent changes to the El Nino circulation
→ retreat of glaciers and thinning of ice sheets
→ changes in the distribution of climates
→ sea levels rising.

Economic changes

Some areas will benefit from global warming. For example, it is suggested that parts of North Africa and the Sahara will become warmer, boosting farming in those areas. In contrast, in areas that become colder and/or drier, farming will be less easy, and there may be increased costs of heating and irrigation. Increased heating costs increase the demand for oil, and so the price of oil may rise. That would have a major impact on all aspects of the economy.

Social changes

Some changes may be beneficial – the growth in street cafés and dining outside in Britain for example. On the other hand, water shortages in Northern Kenya have led to fighting between rival tribes and deaths of tribesmen. Changes in local bye-laws regarding watering of gardens, washing of cars etc. could be some of the less significant effects of climate change.

Political change

Political changes may occur at a local level – how water resources are controlled and the development of water resources in international drainage basins are good examples. The Mekong Drainage basin is one which is set to be developed by China, Laos, Vietnam and Cambodia.

Global dimming

The theory of **global dimming** suggests that the amount of sunlight falling on the Earth's surface is declining. Now, two groups of scientists believe that global dimming came to a halt in the 1990s. In other words, it is now getting sunnier.

According to global dimming, the amount of sunlight reaching the earth in the early 1980s was ten per cent lower than in the 1960s. Further analysis suggested that the global fall was actually slightly less than this (between 7% and 9%), while in some places, such as the Soviet Union, the drop was higher (in that particular case, 20%).

Checkpoint 1

What is global dimming?

Records of sunlight energy received at the earth's surface between 1983 and 2001 suggest it declined until about 1990, and then started rising. Overall, it increased by about 2% over the study period.

The reason is thought to be a decrease in the Earth's albedo from 2000 to 2004, a finding that contradicts earlier results. Much of the sunlight reflected into space by the Earth is turned away by clouds and dust in the atmosphere.

In 2004, a group of researchers suggested that global dimming was caused by clouds getting clogged up with particulate matter such as soot. These particles act as nuclei on which water vapour condenses to form clouds. When such nuclei are rare, clouds tend to be formed of relatively few, large droplets. When they are common, many small droplets form and lots of small droplets are more reflective than a few large ones.

The change from dimming to brightening occurs when the atmosphere is cloud-free. This shift was due to a decrease in the amount of particulate material in the atmosphere. And that, in turn, was due to more effective clean-air regulations and a decline in the economies of east European countries in the late 1980s. The pattern of brightening was also found in other parts of the world, such as North America, Antarctica, Japan and Australia. In rapidly industrialising (and dusty) India, however, dimming continued. Not surprisingly, during the period of global dimming the dirty heavy industry of the Soviet Union dimmed that country most of all.

The jargon

The term *albedo* means reflectivity – it measures the amount of solar radiation reflected by a surface. Light surfaces are more reflective than dark surfaces.

Checkpoint 2

What are the causes of global dimming?

Grade booster

Use up-to-date information, and also indicate that there is more than one aspect to the issue. For example, as well as the impacts of global warming, also consider the impacts of global dimming. Scientists do not fully understand the way that climate works and is changing.

Exam practice   answers: page 30

What are the issues related to climate change? (20 mins)

Coping with climate change

Check the net

Visit the Met Office's Home Page at www.metoffice.gov.uk to find out what is happening about climate change both globally and in the UK.

'The ultimate objective is to achieve ... stabilization of greenhouse gas concentrations in the atmosphere at a level that would prevent dangerous anthropogenic interference with the climate system.'

Checkpoint 1

Why have some countries found it difficult to reach the Kyoto targets?

Climate change is affecting all nations, either directly or indirectly. The causes are varied and the sources are global. As a result, to tackle global climate change there need to be policies that include most nations to work together to achieve a common goal. Making this work is no easy task.

The Montreal Protocol, 1987

The first world conference on climate change was held in Geneva in 1979. However, the Montreal Protocol was the crucial first step in limiting further damage to the ozone layer in the stratosphere. It was signed by many countries to greatly reduce the production and use of CFCs that had been shown to be responsible for damage to the ozone layer. The main CFCs have not been produced by any of the signatories since 1995, except for a limited amount for essential uses, such as for medical sprays.

The Toronto Conference, 1988

This called for the reduction of CO_2 emissions by 20% of the 1988 levels by 2005. Also in 1988 UNEP and the World Meteorological Organisation established the Intergovernmental Panel on Climate Change (IPCC).

The Kyoto Protocol, 1997

This gave all MEDCs legally binding targets for cuts in emissions from the 1990 level by 2008–2012. The EU agreed to cut emissions by 8%, Japan 7% and the USA by 6%. Some countries found it easier to make cuts than others. The UK, for example, had already reduced its use of coal and oil and changed to natural gas. Germany, too, had exhausted many of its traditional heavy industries – such as iron and steel that consumes vast quantities of coal – and Japan was largely dependent on nuclear power and hydroelectric power. By contrast, the USA was experiencing an economic boom in the 1990s and was reluctant to make cuts. According to Kyoto MEDCs were on average to make cuts by 5.2%.

Bali Conference, 2007

Here, all countries, including the US, made a commitment to make 'deep cuts' in greenhouse gas emissions. Two years of negotiations will now start. Rich countries have agreed that poor countries must be given money to help them adapt to climate change. Money may also go to countries not to cut down or degrade forests – one of the most serious sources of climate change emissions. Developing countries will be helped to cut emissions with a transfer of new technologies. Small-scale projects intended to cut emissions will be helped more. The least developed countries will be given extra help to adapt.

However, there are no clear goals or timetables set for emission reductions, only vague guidelines that 'deep cuts' should be made. These could be watered down.

No significant extra money was pledged until after 2012 to help poor countries adapt. No binding targets were set for future funding. It is expected to cost poor countries in the region of £25bn a year to adapt to climate change. There is a danger that climate change money will not go to benefit local people and could even be taken from existing development aid budgets. Extra forestry money could be hijacked by industry for plantations.

Checkpoint 2

In what ways have the methods to tackle climate change differed between the Kyoto Conference and the Bali Conference?

Grade booster

Don't be afraid to tackle controversy – an issue as big as climate change will never be 'black and white'. Indicate how different people, nations and TNCs may differ in their views regarding the causes, consequences and solutions to the issues.

Exam practice answers: pages 30–31

In what ways is it possible to manage climate change at an international level? (20 mins)

Microclimates

Don't forget!

Urban climates are only noticeable during periods of high pressure. During times of low pressure, winds mix the air above urban areas with that of surrounding rural areas.

The jargon

Albedo means the reflectivity of a surface.

Checkpoint 1

Why is temperature higher in urban areas than in rural areas?

Checkpoint 2

Why is thunder more common in urban areas?

Check the net

Visit the Met Office website on Microclimates at: www.metoffice.gov.uk/education/secondary/students/microclimates.html

Urban climates in temperate latitudes differ quite distinctively from nearby rural ones. This is due to a number of reasons.

Structure of the air above the urban area

→ More dust in urban areas means an increased concentration of **hygroscopic particles**.
→ Less water vapour.
→ More CO_2.
→ Higher proportions of noxious fumes from the combustion of fuels.
→ Waste gases discharged by industry.

Structure of the urban surface

→ More heat-retaining materials with lower albedo and better radiation-absorbing properties.
→ Rougher surfaces with a great variety of perpendicular slopes facing different aspects.
→ Tall buildings can be very exposed.
→ Deep streets are sheltered and shaded.

Resultant processes

1 Radiation and sunshine

There is greater scattering of **shorter-wave radiation** by dust, but much higher absorption of longer waves owing to building surfaces and CO_2. Hence there is more diffuse radiation, with considerable local contrasts, because of variable screening by tall buildings in narrow, shaded streets. Reduced visibility arises from **industrial haze**.

2 Clouds and fog

There is a higher incidence of thicker cloud cover in summer and radiation fogs or **smogs** in winter because of increased convection and air pollution. Hygroscopic particles accelerate onset of condensation.

3 Temperatures

Greater heat energy retention and release, including fuel combustion, gives significant temperature increases from suburbs into the centre of built-up areas, creating **heat islands** (see opposite). These can be up to 4°C warmer during winter nights. Snow in rural areas increases its albedo, thereby increasing the differences. Heating from below increases air mass instability overhead, notably during summer afternoons and evenings. There can be strong local contrasts between sunny and shaded surfaces, especially in the spring.

4 Pressure and winds

Severe **gusting and turbulence** around tall buildings causes local pressure gradients between windward to leeward walls. Deep, narrow streets can be calmer unless aligned with prevailing winds to funnel flows along them – the **canyon effect**.

5 Humidity

Decreases in relative humidity occur in inner cities owing to lack of available moisture and higher temperatures. Partly countered in very cold, stable conditions by early onset of condensation in industrial zones.

6 Precipitation

In urban areas there can be more intense storms, because of greater **instability** and stronger convection above built-up areas. Snowfalls are lighter and briefer in urban areas.

The urban heat island

Temperatures are 2–4°C higher in urban areas, creating an urban heat island. This data shows average changes in climate caused by urbanisation:

Factor	Comparison with rural environments	
Radiation	In general	2–10% less
	Ultraviolet, winter	30% less
	Ultraviolet, summer	5% less
	Sunshine duration	5–15% less
Temperature	Annual mean	1°C more
	Sunshine days	2–6°C more
	Greatest difference at night	11°C more
	Winter maximum	1.5°C more
Frost-free season		2–3 weeks more
Wind speed	Annual mean	10–20% less
	Gusts	10–20% less
	Calms	5–20% more
Relative humidity	Winter	2% less
	Summer	8–10% less
Precipitation	Total	5–30% more
	Number of rain days	10% more
Snow days		14% less
Cloudiness	Cover	5–10% more
	Fog, winter	100% more
	Fog, summer	30% more
	Condensation nuclei	10 times more
Gases		5–25 times more

Exam practice   answer: page 31

Describe and explain the main characteristics of an urban climate. (30 mins)

Checkpoint 3

Explain why urban areas are warmer than nearby rural areas, especially at dawn during high pressure conditions?

Checkpoint 4

Why is snow rare in urban areas in the UK?

Examiner's secrets

Try to use local information. Your local council may have some data. Even the climate of the school grounds makes a useful source of data on urban climates.

Answers
Climate

Weather and climate

Checkpoints

1 'Monsoon' is derived from the Arabic word *mausim* meaning seasons, and refers to a seasonal wind. It is also used to describe the rains that come with the moist winds.

2 This is a breeze that blows from land to sea at night and then reverses to blow from sea to land during the day.

3 The jet stream is a high-altitude air movement from west to east.

Exam practice

This essay requires a clear distinction between the winter monsoon and the summer monsoon. This is best done with the use of a climate graph and then a clear description and explanation of the seasonal differences. The main points to remember are as follows. The monsoon is the reversal of pressure and winds, which gives rise to a marked seasonality of rainfall over north and south-east Asia. In winter, temperatures over central Asia are low, leading to high pressure. This brings outward-blowing north-easterly winds across south Asia. These dry air streams produce clear skies and sunny weather over most of India (November–May). Consequently there is very little rainfall.

By contrast, in summer an intense low pressure area develops over Asia. Strong convectional heating under clear skies also leads to low pressure drawing in warm, moist air from over the Indian Ocean. Winds are drawn into the Asian low pressure area bringing with them heavy rain. For example, Cherrapunji can receive over 10 000 mm in just four months.

Climate change

Checkpoints

1 Methane and especially CFCs have a much greater global warming potential than the same volume of CO_2.

2 Ultraviolet radiation breaks down ozone and oxygen into individual atoms. These atoms can then combine to form ozone (O_3).

Exam practice

1 The 'greenhouse effect' is the process by which certain gases, such as CO_2, methane and CFCs, absorb outgoing long-wave radiation from the earth, and return some of it back to earth, thereby raising the earth's temperature.

2 With increased atmospheric energy there is greater potential for more intense storms (more energy means bigger storms) as well as more frequent storms.

Issues of climate change

Checkpoints

1 Global dimming suggests that the amount of sunlight falling on the Earth's surface is declining. The amount of sunlight reaching the earth in the early 1980s was 10% lower than in the 1960s.

2 Global dimming is caused by clouds getting clogged up with particulate matter such as soot. These particles act as nuclei on which water vapour condenses to form clouds. When such nuclei are rare, clouds tend to be formed of relatively few, large droplets. When they are common, many small droplets form and lots of small droplets are more reflective than a few large ones.

Exam practice

Environmental issues include increased flood frequency and flood magnitude; more frequent changes to the El Nino circulation; retreat of glaciers and thinning of ice sheets; changes in the distribution of climates; rising sea levels.

Economic changes are varied. Some areas will benefit from global warming. For example, it is suggested that parts of North Africa and the Sahara will become warmer boosting farming in those areas. In contrast, in areas that become colder and/or drier, farming will be less easy, and there may be increased costs of heating and irrigation. Social changes may be beneficial or negative – the growth in street cafés and dining outside in Britain for example. On the other hand, water shortages in Northern Kenya have led to fighting between rival tribes and deaths of tribesmen. Changes in local bye-laws regarding watering of gardens, washing of cars etc. may appear to be some of the petty effects of climate change. Political changes may occur at a local level – how water resources are controlled and the development of water resources in international drainage basins are good examples.

Another issue is whether global warming is the cause of change to global dimming. Global dimming suggests that the amount of sunlight falling on the Earth's surface is declining. Records of sunlight energy received at the earth's surface between 1983 and 2001 suggest it declined until about 1990, and then started rising. Overall, it increased by about 2% over the study period.

Coping with climate change

Checkpoints

1 Many NICs find it difficult to agree to the Kyoto targets as it would limit their potential for economic development. Similarly, some MEDCs, notably the USA, do not want to reduce carbon emissions, as it would harm their economy, and make them less powerful than some of the emerging NICs, such as India and China.

2 The Kyoto Protocol gave MEDCs legally binding targets for cuts in emissions from the 1990 level by 2008–2012. However, a number of countries (notably the USA) did not sign up for the Kyoto Protocol, and Russia and

Australia were late signatories. However, at the Bali Conference in 2007, all countries, including the USA, made a commitment to make 'deep cuts' in greenhouse gas emissions. However, there were no clear goals or timetables set for emission reductions, only guidelines that 'deep cuts' should be made.

Exam practice

You should show an awareness of the methods that could reduce the impacts of climate change, but also show an awareness that it has not proved possible to achieve these goals.

It is proving difficult to manage climate change at an international scale. Although the principles are well known – for example, reduce emissions of CO_2 (as per the Kyoto Protocol), decreases rates of deforestation, use less fossil fuels and so on – the real difficulty is getting countries to agree to do such actions, and then to enforce them.

For examples, NICs such as China and India argue that they should be allowed to industrialise and develop, as countries in the West have done. To achieve this requires the burning of large amounts of fossil fuels. Other countries, notably the USA, have resisted reductions in their emissions of carbon, arguing that there is insufficient evidence linking climate change to human activities. Some lobbyists argue that there is too much at stake for the economy if developed countries reduce their carbon emissions.

Microclimates

Checkpoints

1 A city heat island is caused by both the structure of the city and the different air above it. The city also generates heat from traffic and buildings, which can get trapped.
2 Air is more unstable above the city and the strong convection surges can trigger thunderstorms.
3 In the rural area the clear sky brought by high pressure allows rapid cooling at night, whereas the city traps the residual heat because of the buildings.
4 The urban heat island causes precipitation to fall as rain rather than snow. If snow does fall it melts rapidly.

Exam practice

This is a straightforward essay in two parts. The first part should describe succinctly the main characteristics of an urban climate, such as: temperature and heat island effect; amount and type of precipitation; wind speed and wind characteristics; humidity; and cloud cover. The explanation should include: sources of heat from industry and offices; albedo; specific heat capacity of buildings; absence of vegetation and water bodies; and the impact of homes, cars and other forms of transport.

Top answers will include lots of examples, many from local studies – even from the microclimate of the school.

Revision checklist
Climate

1	Describe the main differences between weather and climate.	Confident	Not confident **Revise** page 20
2	Describe the main characteristics of hot desert climate.	Confident	Not confident **Revise** page 20
3	Describe and explain the main characteristics of the monsoon.	Confident	Not confident **Revise** pages 20–21
4	Describe the natural and man-made causes of climate change.	Confident	Not confident **Revise** pages 22–23
5	Illustrate the issues connected to climate change.	Confident	Not confident **Revise** pages 24–25
6	Explain ways of coping with climate change.	Confident	Not confident **Revise** pages 26–27
7	Describe and account for the main features of microclimates.	Confident	Not confident **Revise** pages 28–29

Water

The study of the processes which shape our river channels and river basins is an important part of some Geography specifications at AS level. This involves the study of the physical processes of erosion, transport and deposition as they occur in river basins and recognizing landforms that are produced as a result. It also involves the study of how river basins change over time and the variety of factors that go to shape any river basin. River discharge is frequently illustrated using a type of graph called a storm hydrograph and it is important that you understand what these show and can recognize the discharge patterns illustrated by contrasting examples.

Exam themes

- Storm hydrographs
- River channel processes
- Base level and graded river profiles
- Features of upper, middle and lower valleys

Topic checklist

	Edexcel		AQA		OCR		WJEC	
	AS	A2	AS	A2	AS	A2	AS	A2
Hydrological processes			O		O		O	
Rivers and their landforms			O		O		O	

Hydrological processes

The jargon

A *drainage basin* is an area of land bounded by a watershed, within which all surface and subsurface water will eventually find its way into the same river.

Action point

Draw a systems diagram of a drainage basin showing the inputs, stores, processes (or transfers), and outputs listed in the table.

Action point

Make a list of the features of the storm hydrograph that are labelled on this diagram, and explain the meaning of each feature.

Action point

All drainage basins are different but you should be able to explain how a variation in each of the variables listed here will affect the shape of the storm hydrograph.

Checkpoint 1

Describe the characteristics of particles carried by each of the four processes of transportation.

Running water shapes the landscape over which it flows in a number of ways. These processes operate within a series of drainage basins or catchment areas that supply the river systems with water. The processes operating in any drainage basin are related to the discharge of the river, which varies over time.

Storm hydrographs

A hydrograph shows changes in river discharge over a period of time. A storm hydrograph shows the way that the discharge at one point along a river varies following a storm event in the catchment area.

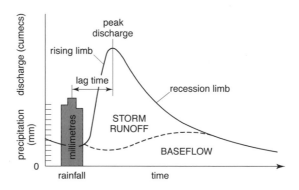

The way a drainage basin responds to a storm event can vary according to:

→ the size of the basin
→ the shape of the basin
→ the drainage density within the basin
→ the porosity and permeability of the soils
→ the geology underlying the basin
→ the steepness of the slopes in the basin
→ the type of vegetation cover in the basin
→ the type of land use, particularly the urban areas within the basin
→ the level of residual moisture in the soil layer
→ the amount, intensity and duration of the rainfall
→ the seasonal weather conditions.

River channel processes

Rivers will try to adopt a channel and basin shape that allows them to fulfil their role of transporting water and sediment most efficiently. They do this through the following processes:

Transportation

Rivers transport materials by **traction**, **saltation**, **suspension** and **solution**. Most rivers carry about three-quarters of their load as suspended sediments, depending upon the local rock type, the climate, and the velocity of flow. There may also be seasonal variations.

Erosion

Erosion is the series of processes by which a river wears away its banks and its bed. There are four different processes:

→ **abrasion, corrasion** – similar to the action of sandpaper

→ **corrosion/solution** – a chemical process affecting certain rocks
→ **hydraulic action** – the force of water moving loose material and compressing the air in cracks in the banks, and loosening material
→ **attrition** – the breaking down and smoothing of particles as they are rubbed and knocked against each other.

Deposition

As a river slows down it loses energy, and thus its **competence** or **capacity** to transport its load is reduced. Deposition happens when there is a period of low discharge or velocity of flow falls as the river enters a lake or the sea. Material is also deposited when a river overflows its banks and its velocity falls, and when shallow water is encountered, e.g. on the inside of a meander bend. In addition, when the load is suddenly increased following a landslide, more material will be deposited downstream.

The Hjulström curve

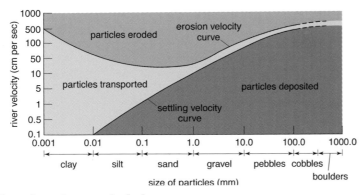

Base level and a graded river

The base level is the lowest level to which a river may erode; for most rivers this is sea level. Temporary or local base levels also occur when a river flows into a lake, or where there is a resistant band of rock across the valley. Base levels may be the result of climatic or tectonic changes.

A positive change is when sea level rises in relation to the land; a negative change is the opposite. If the base level changes in any way due to either **eustatic** or **isostatic changes**, the river will adjust its long profile accordingly in an attempt to achieve a **graded profile**. If the base level falls, the **rejuvenation** process starts at the sea and works its way upstream; the point at which the change is occurring is the **knickpoint**. Rejuvenation also leads to the formation of **river terraces** and **entrenched** or **ingrown meanders**.

A river with a graded state will also have its cross-profile and channel roughness in balance with the discharge and load.

Exam practice   answers: page 38

1 Describe how and explain why the lag time of a river in a catchment area of deciduous woodland would change from summer to winter. (10 mins)
2 Explain how erosion and deposition will tend to create a graded profile in the river shown below. (10 mins)

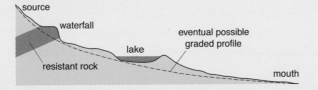

Rivers and their landforms

The processes that are operating in the river channel produce a variety of distinctive landforms. These can be classified according to their location within the basin or to the main processes that are responsible for their formation. Neither classification is perfect as many features can be found in different locations within the basin and many features are not the exclusive result of either erosion or deposition. Here the features are described according to their more common location within the drainage basin.

Features of the upper valley

In the upper reaches of the valley the river channel is usually rocky and full of angular boulders. The river channel is therefore very inefficient and there is little energy to pick up and transport material. When the river discharge increases there is more energy available and, as vertical erosion is the dominant process, a **V-shaped valley** with **interlocking spurs** results. As the turbulent water swirls around, pebbles are responsible for scouring out depressions that develop into **potholes**.

The upper sections of a river also contain:

→ **waterfalls and rapids**
→ **gorges**
→ **riffles and pools**
→ **springs**.

Action point

Can you describe and explain the formation of each of these features?

Features of the middle valley

Meanders are found in all sections of the valley and are a result of the processes of erosion, transportation and deposition within the channel. There is some doubt about the exact processes that form them, but the alternate riffles and pools that are found in relatively straight sections of rivers and **helicoidal flow** are thought to be involved.

The jargon

Helicoidal flow is the corkscrew-like movement of water molecules as they move downstream.

Checkpoint 1

Explain how oxbow lakes or cut-offs are formed.

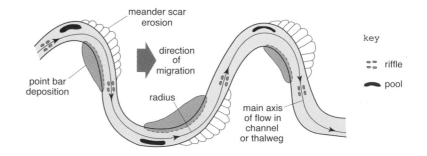

Floodplains are formed beyond the banks of the river by deposition of layers of fine fertile silt at times of flood. As the water rises above **bankfull stage** the hydraulic radius and friction increases resulting in deposition. The floodplain may be widened by lateral erosion of meanders up to the **bluff line**.

Features of the lower valley

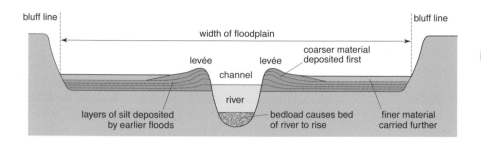

bluff line — width of floodplain — bluff line

levée — levée — coarser material deposited first

channel

river

layers of silt deposited by earlier floods — bedload causes bed of river to rise — finer material carried further

Checkpoint 2

Use this diagram to help you explain the formation of levées and bluffs.

Deltas

As a river nears base level it slows down and deposition becomes the dominant process. The deposited material gradually builds up into a swampy plain known as a **delta**.

The main processes of deposition in this environment are:

→ bedload dumping
→ settlement of suspended sediments
→ **flocculation**.

If the rate of deposition exceeds the rate of marine erosion the delta will continue to form. As the quantities of deposited sediment continue to grow, the channel will become choked and it will be forced to divide into a number of **distributaries** that grow into one of four characteristic shapes:

→ cuspate
→ arcuate
→ bird's-foot or lobate
→ estuarine.

The jargon

Flocculation is the coagulation of clay particles in suspension when they meet sea water causing them to settle on the bed quickly.

Action point

Draw a series of simple diagrams to illustrate the four different types of delta.

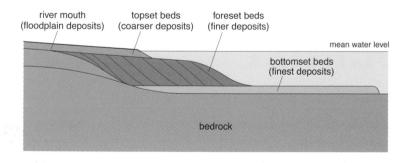

river mouth (floodplain deposits) — topset beds (coarser deposits) — foreset beds (finer deposits)

mean water level

bottomset beds (finest deposits)

bedrock

Links

Managing river basins, page 34.

Grade booster

Examples chosen from your own fieldwork or research are often easier to describe in detail. Try to use personal experience whenever possible to give your answers increased conviction.

Exam practice — answers: page 38

1 How might the following changes in the drainage basin affect the growth of a delta: (a) building a dam, (b) deforestation of several slopes? (10 mins)

2 How might the characteristics of a lower river valley be altered by rejuvenation? (10 mins)

Answers
Water

Hydrological processes

Checkpoints

1 Traction moves large, heavy material along the bed. Saltation is the jumping or bouncing of pebbles, sand and gravel along the bed. Suspension is fine particles of clay and silt carried by turbulence in fast-flowing water. Solution is dissolved material.

2 *Competence* is the diameter of the largest particle that a river is able to carry as it moves at a certain speed. *Capacity* is the largest amount of load that a river can carry as it moves at a certain speed.

Exam practice

1 Due to the leaves on the trees in summer there will be more interception and evaporation, resulting in a much longer lag time. You will need to explain the processes of interception, stemflow and evaporation.

2 The graded profile is a smooth profile attained when the river is in a state of dynamic equilibrium. In the situation shown the lake will gradually fill up as deposition will be greater than erosion in this part of the system. At the waterfall, erosion will be more important than deposition and the feature will eventually retreat and disappear.

Rivers and their landforms

Checkpoints

1 Oxbow lakes are formed as a river becomes increasingly sinuous. Erosion on the outside of the bend and deposition on the inside of the bend lead to the neck of the meander becoming increasingly narrow until the main channel cuts through. The section of the former channel becomes silted up and is called the oxbow lake.

2 Levées are the parts of the riverbank that are higher than the rest of the floodplain. They are built up from layers of sediment dropped when the river floods. The coarsest material is deposited closest to the river. Bluffs are steep cliffs along the edge of the floodplain. They are formed as a result of lateral erosion by the river up to the edge of the floodplain and into the surrounding higher land.

Exam practice

1 (a) Dam construction will reduce the amount of sediment available to build the delta and if erosion by the sea remains constant the delta may well reduce in size.

(b) Deforestation will reduce the vegetation cover and tend to increase the soil erosion leading to an increase in the load of the river. This will cause an increase in the development of the delta if all other factors remain unchanged.

2 Rejuvenation means a lowering of base level and this leads to greater erosion, particularly near the mouth. Your answer needs to contain reference to knick points, river terraces and incised meanders. You should also mention the way that the extra powers of erosion will be transferred upstream as the processes continue.

Revision checklist
Water

By the end of this chapter you should be able to:

1	Describe the various parts of a storm hydrograph.	Confident	Not confident **Revise** pages 34–35
2	Explain how the storm hydrograph can be used to show the way a river discharge relates to a storm event.	Confident	Not confident **Revise** pages 34–35
3	Describe the main processes of fluvial erosion, depositions and transportation.	Confident	Not confident **Revise** pages 34–35
4	Describe and explain how rivers achieve a graded profile.	Confident	Not confident **Revise** pages 34–35
5	Describe and explain the main landforms created by fluvial processes in the upper, middle and lower parts of a river basin.	Confident	Not confident **Revise** pages 36–37

Biomes and ecosystems

The study of ecosystems is a key element of geography. Within physical geography the study of the nature, structure and processes in ecosystems, together with patterns of their distribution, is the field of biogeography. Understanding ecosystems, though, is also essential in any study of the interaction of people and their environments, for we need to understand what effect people's actions may have on ecosystems. We also need to understand how ecosystems can provide resources for human use. The field of sustainable development is underpinned, therefore, by a knowledge of how ecosystems work.

Exam themes

- Tropical forest and grassland ecosystems
- Ecosystem processes and structures
- Small-scale ecosystems
- Ecological succession
- Threats to biodiversity

Topic checklist

	EDEXCEL		AQA		OCR		WJEC	
	AS	A2	AS	A2	AS	A2	AS	A2
Tropical forest and grassland ecosystems				●				
Ecosystem processes and structures		●		●		●		●
Small-scale ecosystems				●		●		●
Ecological succession		●		●		●		●
Threats to biodiversity		●		●		●		●

Tropical forest and grassland ecosystems

Forests originally covered 60% of the earth's surface, while grassland covered 20%. Today these proportions are 30% and 40% respectively as a result of forest clearance.

Tropical broad-leaved forest

Tropical broad-leaved forest (monsoon forest or tropical rainforest) is found 5 degrees either side of the Equator where temperatures and rainfall are high throughout the year. Rich in biodiversity (up to 60 species of plants and animals per hectare), these forests have three distinct layers – ground layer with few plants, the middle layer with most species, and the canopy layer of the largest trees.

The clearance of the tropical broad-leaved forests has been an issue of major international concern, for some 20 million hectares per annum are lost through logging. The reasons for clearance are:

→ demand for tropical hardwoods (e.g. mahogany) from MEDCs
→ pressure to clear land for agriculture in LEDCs:
    → to meet the needs of a rapidly growing population
    → to produce export agricultural goods (e.g. beef)
→ economic pressure on LEDCs to export valuable forest products
→ for firewood at the edges of forests in densely populated regions.

The environmental effects of forest clearance include:

→ the loss of trees that play a major role in absorbing atmospheric carbon dioxide and releasing oxygen
→ a rapid loss of fertility in soils:
    → most of the nutrients in the ecosystem are located in the trees, and are lost with them
    → the exposed soils are leached by the heavy rainfall
→ soil erosion, because of the heavy rainfall on exposed soils
→ flooding, resulting from the erosion of soils into the rivers
→ loss of species because of the loss of habitat – tropical forests contain large numbers of species, many unknown, which might be a resource, e.g. for future pharmaceuticals.

The strategies used to reduce forest clearance and its impacts are:

→ decreasing demand for tropical forest products such as mahogany
→ encouraging tropical countries to export much higher-value processed timber (e.g. plywood) rather than logs
→ stricter management of logging to reduce wastage, encourage replanting and reduce erosion after clearance
→ using alternative fuel sources in LEDCs, such as biogas digesters, mini-HEP schemes, or developing fuelwood plantations.

Checkpoint 1

Check the temperature and precipitation conditions under which tropical forest will be found.

Action point

Choose a country as a case study of tropical forest management. Brazil, Indonesia and Malaysia are amongst the best examples.

Checkpoint 2

Reducing the loss of species is called 'maintaining biodiversity'. Why is this important to people?

Check the net

Identify what Greenpeace says about forest management at: www.greenpeace.org

Tropical grassland ecosystems

Tropical grasslands, known as **savannas** in Africa and **llanos** in South America, lie between the tropical forests and the hot deserts, typically between 10 and 17 degrees north and south of the Equator. The climate here has distinct wet and dry seasons, with wet seasons occurring as the Inter-tropical convergence zone (ITCZ) moves with the overhead sun.

Seasonal rainfall is insufficient to support forest, except in some river valleys, but grasses flourish during wet seasons and survive dry seasons. In wetter zones scrub or open woodland (e.g. acacia thorn trees) survive. Grasses are supported by nutrients from frequent grassland fires during the dry season, often triggered by lightning. Much of this zone has been converted to farmland, but in biosphere reserves and estates the grasslands support a wide range of herbivores (e.g. zebra, wildebeest) and carnivores (e.g. lion, leopard), each with a distinctive ecological niche – for example, in southern Africa white rhino graze grasses while black rhino eat taller shrubs and bush leaves, while giraffes eat thorn bush leaves at high levels.

Key environmental issues are:
→ pressure to develop farmland because of rising populations, for example in many areas of Kenya and Tanzania farmland now dominates the savanna areas
→ the creation of biosphere reserves (e.g. game parks). These both preserve the savanna ecosystem, and provide a valuable economic benefit through tourism. Shamwari Game Reserve in Eastern Cape, South Africa, has been created by returning marginal farmland to its natural ecosystem, and now provides income and employment through tourism.
→ pressure on soil and vegetation in game parks from increasing animal numbers and from tourism. Protection of species can lead to overpopulation, which in turn can lead to habitat destruction and then to soil erosion.

Strategies to deal with these issues include the culling of animals to maintain numbers within carrying capacities, and the zoning of reserves to focus tourism in some areas and protect others for wildlife. Shamwari Game Reserve uses 'excess' animals as a cash crop by selling on wild animals to other game reserves.

Checkpoint 3

What are the main countries from which tourists visit the game reserves of East Africa? Think about where responsibility lies for erosion by tourists.

Test yourself

Draw a flow diagram to show the stages/steps that link the placing of animals in game reserves and the start of soil erosion.

Checkpoint 4

List three advantages and three disadvantages of biosphere reserves.

Action point

Find out the difference between sheet erosion and gully erosion.

Grade booster

Try to use some up-to-date examples that are currently topical and which have been in the news or covered in sources such as 'Geographical' magazine very recently.

Exam practice   answers: page 50

1   With reference to a specific case study, discuss the environmental, social and economic issues resulting from tropical forest clearance. (12 mins)

2   Critically evaluate the management strategies used to reduce the environmental impact of game reserves in savanna regions. (12 mins)

Examiner's secrets

Use a quick sketch map to locate a case study. Examiners want you to show that you can argue a case as well as know the facts, so you need to show that you do or do not agree with particular statements.

Ecosystem processes and structures

An ecosystem comprises the plants, animals and non-living components (e.g. water) that exist together in a particular location and also the processes (e.g. photosynthesis) that operate in that environment. They can be any size, from a puddle, to a lake, to one of the world's large natural vegetation zones (biomes), such as the Amazon tropical rainforest.

Checkpoint 1

What would be the main components of a pond ecosystem? Don't forget living (biotic) and non-living (abiotic) things. Try to distinguish inputs, outputs and processes in the pond ecosystem.

The jargon

Gross primary productivity (GPP) is the amount of energy photosynthesised. *Net primary productivity* (NPP) is GPP minus the energy used for respiration by the plant.

Energy flow in ecosystems

All the energy required for the processes in ecosystems is obtained from the sun. This energy is transferred through the ecosystem as follows:

→ Energy from sunlight is captured by green plants and converted into plant material and sugars by photosynthesis. The sugars are used by the plant for living processes, but some energy becomes plant material (**biomass**), an amount called the **net primary productivity (NPP)**.

→ Plants are eaten by **herbivores**, which gain their energy for life from the plant material, but also store some of the energy as biomass.

→ Herbivores are eaten by **carnivores**, who use some of the energy for living (reproduction, movement, etc.), but convert some to biomass.

→ Carnivores may be eaten by **top carnivores**.

An example of such a **food chain** is: leaves → ants → voles → hawks. In practice, most animals feed off several species, so ecosystems are really **food webs**.

Checkpoint 2

Write a simple definition of the terms highlighted in **bold** in this section.

Action point

Draw a food web for an ecosystem you have studied.

Checkpoint 3

Try to explain why there may be only one tiger in a large area of tropical forest in northern India.

Trophic levels and ecological pyramids

The figure below shows an ecological pyramid representing organisms in a pond. Each level in the pyramid is called a trophic level. The number of individuals and the amount of biomass decrease from level to level as there is less energy available to support life.

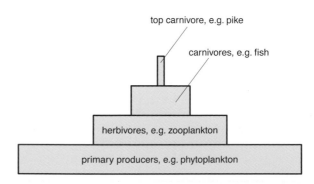

top carnivore, e.g. pike
carnivores, e.g. fish
herbivores, e.g. zooplankton
primary producers, e.g. phytoplankton

Productivity in ecosystems

The world's major ecosystems vary in productivity. High productivity is found where:

→ there are high temperatures (e.g. in the tropics)
→ there is a plentiful supply of water (either as precipitation or in a water ecosystem such as a lake)
→ there is a plentiful supply of nutrients (e.g. where rivers carry nutrients into an estuary, or where ocean currents converge).

High NPP is found, for example, in a tropical rainforest (2000 g/m²/year). Low NPP is found in hot deserts or tundra (e.g. 100 g/m²/year).

Nutrient cycles

Nutrients such as carbon, nitrogen or potassium are the chemical elements that are essential to support life. Plants and animals obtain them through the food chain. Nutrient cycles show the pathways that nutrients follow within the environment. The figure below shows a simple model of the carbon nutrient cycle. Within the cycle:

→ carbon occurs in different forms, e.g. as carbon dioxide gas in the air, as plant sugars, or as limestone rock
→ large amounts of carbon are stored for long periods as rocks
→ there is human interference, e.g. releasing more carbon dioxide into the air by burning fossil fuels or clearing forests.

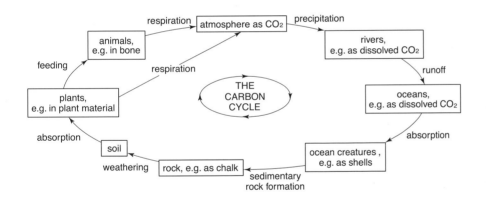

Action point

For one of the processes within the carbon cycle, investigate from textbooks how that process operates.

Checkpoint 4

What are the main ways people interfere with the carbon cycle?

Grade booster

Learn and remember the details of productivity in specific ecosystems e.g. contrasting tundra ecosystems with tropical forest ecosystems.

Examiner's secrets

Maximum marks require you to use named examples of ecosystems or real cases of ecosystem management issues. Annotated diagrams can save you a lot of writing.

Exam practice   answers: page 50

1   The table below shows the net primary productivity (g/m²/year) for a number of major biomes.
    (a) Explain the concept of net primary productivity. (5 mins)
    (b) With reference to specific examples, explain why NPP varies between biomes. (8 mins)

Biome	Mean net primary productivity
Tundra	140
Estuaries	2000
Temperate grassland	600
Temperate deciduous forest	1200
Tropical evergreen forest	2200

2   With reference to the carbon cycle, show how human activity can have an impact on natural nutrient cycles. (15 mins)

Small-scale ecosystems

Understanding the effects of human interference in the environment means it is important to understand ecosystems at a small scale. Small scale ecosystems include, for example, a local pond or river, or an area of woodland. Most small-scale ecosystems in the UK are the result of considerable human intervention. Urban areas may contain, for example, parks, suburban gardens, derelict industrial areas, roadside verges or disused railway lines. Rural areas will include farmland ecosystems, managed coniferous woodlands or formal parks and estates. Small-scale ecosystems are a good topic for study through geographical fieldwork.

Checkpoint 1

Ecosystems have both biotic and abiotic elements. What is meant by each of these terms.

Components of a small-scale ecosystem

Small scale ecosystems will contain primary producers (green plants), herbivores, and carnivores, whose relationship can be shown as a food web. The diagram below shows a **food web** for a deciduous woodland in southern England.

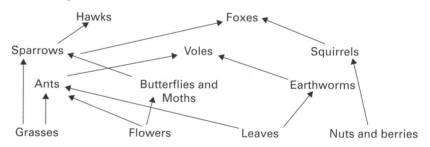

Simple food in a UK woodland

Action point

Choose an example of a small-scale ecosystem that you have studied. Draw a simple food web for the ecosystem and identify the species in each trophic level.

Check the net

Visit the website of your local Council (County Council or District Council). Use this to identify some examples of small-scale ecosystems in your locality that are being protected or managed.

Ecological niches and predator–prey relationships

Each plant or animal in the ecosystem has a unique role and pattern of life. This is its **ecological niche**. In a lake, for example, a pike will be the single top carnivore, and in a tropical savanna each herbivore (e.g. elephant, giraffe) survives by feeding off different plants or parts of plants. In British coastal environments two similar seabirds coexist by living off slightly different food sources – the shag feeds off eels and free-swimming fish, while the cormorant feeds off bottom-living fish and invertebrates. Although each ecological niche is unique, the number of individual animals or plants in that niche will vary from time to time because of the availability of food, as environmental conditions change from year to year. A warm summer may lead to lake plants flourishing, providing plentiful food for insects, which then thrive and provide plentiful food for fish, which in turn are prey for the pike. A cold summer may have the opposite effect. There is therefore a strong **predator–prey relationship**.

Checkpoint 2

Why might a small local lake have only one or two pike fish in it but many hundreds of smaller fish?

Change and feedback in ecosystems

Ecosystems change over time because of changing environmental conditions, or because of interference by people. Introducing new species into an ecosystem will lead to **competition** with existing species for their ecological niche, and only one species can survive – the grey squirrel, introduced from North America, has caused the loss of red squirrels from most woodlands in the United Kingdom, for example.

Changes in ecosystems produce effects elsewhere in the system through **feedback processes**. There are two forms of feedback:

➜ **Positive feedback** occurs where the change becomes permanent (for example with the introduction of the grey squirrel).
➜ **Negative feedback** occurs where the change has consequences which then 'switch off' the change. In this way an ecosystem is self-regulating and is stable over a period of time. This stability in an ecosystem is known as **homeostasis**. For example, a warm wet year leads to plentiful berries for birds, who thrive and produce many young. However, competition between the young birds in the following winter will mean only the fittest survive, and numbers return to the original level.

Small ecosystems often show rapid change in response to changing environmental conditions, as the food webs are simple and each species has only one or two potential food sources. For example, in Arctic ecosystems the lynx feeds largely on snowshoe hares, so a change in hare numbers has an immediate effect on lynx numbers. In large ecosystems such as savanna grassland the food webs are more complex, so changes may have less immediate impact.

Urban ecosystems will show the impact of human intervention. For example, suburban gardens may contain much larger numbers of birds than rural 'natural' forest, but fewer species. This is because the gardens provide a rich habitat for some species which can therefore grow in number with fewer predators.

Checkpoint 3

Why might a garden bird species experience a population crash in the winter following a warm, wet summer producing plentiful berries and seeds?

Test yourself

Draw a simple diagram to illustrate the idea of negative feedback.

Grade booster

Always use examples from your own fieldwork and if possible be able to describe some of the issues and problems you faced with collecting real data in the field.

Exam practice answers: page 51

1 With reference to a specific small scale ecosystem, explain how negative feedback and positive feedback processes can impact upon the ecosystem. (12 mins)
2 Explain the following ecosystem concepts
(a) ecological niches
(b) predator–prey relationships
(c) competition
(d) homeostasis. (12 mins)

Examiner's secrets

Most concepts and models in the study of ecosystems are best illustrated with a simple diagram. Such diagrams will be given credit by the examiners – and may save you writing a long piece of text.

Ecological succession

All ecosystems change over time. There are small changes from year to year even in well-established ecosystems, perhaps because of slight variations in temperature or rainfall. However, such changes usually disappear as a result of negative feedback processes. Changes over long periods of time are known as **ecological succession**, and may result from natural environmental changes or from the impact of human activity.

Primary succession

Primary succession occurs when a new environment is formed. This could be due to:

→ new land being created by tectonic activity, e.g. a new volcanic island
→ other geomorphic processes creating virgin land, e.g. through the formation of sand dunes by wind.

Colonisation by plants progresses over time from pioneer species such as algae to moss, grasses, flowering plants and trees, which become established as soil forms and becomes richer with humus. Animals move in as the food supply grows. Eventually a mature, stable ecosystem, or **climax community**, will develop. Its nature will depend on factors such as rock type, water supply or the climate. The world's major vegetation zones are normally **climatic climax communities**; for example, coniferous forest (taiga) is the natural ecosystem in many cold temperate regions. The community at each stage of succession is known as a **sere**. Various types of sere occur:

→ hydroseres, where the succession is in water, e.g. in a wetland area
→ xeroseres, where the conditions are dry, e.g. in desert
→ psammoseres, where the succession occurs on sand, e.g. dunes.

Succession in coastal dunes

Sand dunes provide a good example of primary succession. The succession proceeds from bare sand through the growth of coarse succulent grass, such as marram grass, which develops a thick root mat, and low-growing herbs. This stabilises the dune and adds humus. Other species such as shrubs and small sand-tolerant trees then colonise, such as willow and birch. Grasses and weeds (e.g. fescue) then spread to create dune grassland, which is invaded by thistle, ragwort and bracken. The figure at the top of the next page shows a transect through dunes illustrating the succession.

The main pressures on dune ecosystems are from people, through trampling, pollution, inappropriate uses (e.g. cycling), and from grazing. These pressures can be managed by:

→ **estate management** – maintaining and fencing paths, etc.
→ **habitat management** – controlling scrub, or planting marram grass
→ **visitor management** – providing signposts, leaflets and trails.

Action point

Choose a climatic climax community, and find out the details of its plants, animals and structure.

Checkpoint 1

Write a definition of each of the types of *succession* and each of the main types of *sere*.

Action point

Take notes on a case study of coastal dune succession and management. Good examples are Studland Bay, Dorset; Oxwich Burrows, South Wales; and Les Mielles in Jersey.

Test yourself

Draw a cross-section of a dune system, labelling the likely plant species and the main management issues.

Checkpoint 2

Make a list of the main types of natural and human factors that might cause secondary succession to begin.

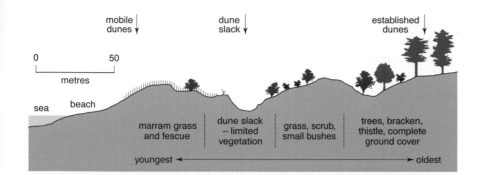

Secondary succession

Secondary succession occurs where a climax community is affected by a permanent environmental change. This could be caused by:

→ a natural change, such as climate change
→ human impact, e.g. through the clearance of woodland for farmland or the draining of a wetland.

Naturally, such a change would produce a new stable climax community. Where people maintain a sub-climax community for their own needs (e.g. farmland, or heather moorland maintained by burning) then a **plagioclimax community** develops.

The New Forest – a plagioclimax community

The New Forest, in Hampshire in southern England, is an example of a managed plagioclimax community. Its mixture of heathland, wetland and open woodland has been maintained for almost a thousand years by grazing, tree planting and drainage. Without active management it would revert to mixed deciduous woodland. Despite being the product of human management, it is now a protected 'natural' landscape.

Checkpoint 3

Write a definition of a *plagioclimax community*.

Grade booster

Research the main ecosystems under threat in your local area and use these as examples of ecological change and succession in your written answers.

Exam practice   answers: page 51

1   The data below show the process of colonisation and succession on the newly created volcanic island of Surtsey, Iceland.

Year	Number of moss species	Number of flowering plant species
1965	5	3
1968	8	6
1971	39	8
1974	65	15

(a) Explain the process shown by these data.

(b) With reference to specific examples, show (i) how succession occurs in a sand dune environment, and (ii) the management issues that result from this process. (15 mins)

2   'Without careful management most "natural" British plant and animal communities would not exist.' With reference to specific examples, explain this statement. (15 mins)

Examiner's secrets

Always quote data from the table supplied as part of your answer.

Threats to biodiversity

Biodiversity is the variety of living things on Earth, ranging from viruses to the largest mammals, and from plankton to forest trees. Although there are 1.5 million identified species of plants and animals, this is believed to represent less than 20% of all the Earth's species. Three levels of biodiversity are recognized:

→ ecosystem diversity, referring to the range of habitats and ecosystems
→ species diversity, referring to the range of species that exist
→ genetic diversity, referring to the diversity and variation *within* a single species.

Checkpoint 1

Explain the difference between species diversity and genetic diversity.

Changes in biodiversity

Biodiversity is the result of evolution since the first life appeared on Earth. Evolution and extinction of species have always occurred, but the greatest biodiversity was probably in the period just before the appearance of humans. There is now concern about rapid decline in biodiversity. Estimates of the number of species lost range from 30,000 to 150,000 per year. The IUCN (International Union for the Conservation of Nature) Red List of Threatened Species has 41,400 species on it, with 16,000 threatened with extinction in 2007.

Check the net

Visit the IUCN website and look at the details of the IUCN Red List (www.iucn.org).

An example of declining biodiversity is the decline of the great apes. The Western Gorilla, which lives in West Africa, is now critically endangered as its population has declined 60% in 25 years, mainly because of hunting for bushmeat. The Sumatran Orangutan is endangered as a result of habitat destruction because of logging and the replanting of habitats with oil palm plantations.

The value of biodiversity

Biodiversity is important to humans because:

Checkpoint 2

Explain why maintaining biodiversity is important to people.

a) Plants and animals are the basis of almost all food and pharmaceuticals and many industrial products
b) Future resource risk. 95% of global food comes from 30 species, which makes humans at high risk from future plant disease
c) Food and drug potential. Future valuable food or pharmaceuticals may be discovered using species not yet recognized as important, or not yet discovered
d) Ecosystem value. Ecosystems have both material benefits (e.g. removing CO_2 from the atmosphere, controlling flooding, or their use for recreation), and non-material benefits (e.g. their aesthetic or spiritual value)
e) Many feel there is a moral obligation on people of stewardship of the Earth and its diversity of life

Check the net

Identify what Greenpeace says about threats to biodiversity at: www.greenpeace.org

Threats to biodiversity

The main threats to biodiversity are:

→ Destruction of habitats. For example, forest cover has declined from 60% to 30% of the land surface since 1600 because of the spread of agriculture and the clearance of forests for timber
→ Overhunting of animals (e.g. tigers) or over-clearance of plants
→ Introduction of exotic species, which then destroy endemic species e.g. the introduction of cats to oceanic islands
→ Pollution
→ Hybridization. Diversity of species and genetic diversity has declined by the breeding of a very small number of specialized food and resource plant and animal species.

Checkpoint 3

What are the main reasons for habitat destruction and overhunting of species?

Managing biodiversity

International efforts to preserve biodiversity have included:

a) The World Conservation Strategy, 1980, developed by the UN Environment Programme (UNEP), the Worldwide Fund for Nature (WWF) and IUCN
b) The UN Framework Convention on Biodiversity, signed at the 1992 Rio Earth Summit
c) International agreements (conventions) e.g CITES , the Convention on International Trade in Endangered Species signed by 110 countries in 1973.

Action point

Visit the website of the government department DEFRA (www.defra.gov. uk). Identify the main strategies being supported by DEFRA to preserve biodiversity in the UK.

Most countries have national programmes to tackle reductions in biodiversity. For example, in the UK, the Biodiversity Action Plan (BAP) is managed by a wide range of charitable bodies, NGOs and the government department DEFRA (Department for Environment, Food and Rural Affairs). Specific management strategies include:

→ Biosphere reserves e.g. national parks and reserves
→ Zoos, aquaria and botanical gardens
→ Gene banks
→ Research programmes.

Test yourself

What are the advantages and disadvantages of biosphere reserves, zoos and gene banks?

Examiner's secrets

You always need to show that you know some detailed examples as well as general principles. Ensure that you have an example or case study to use to illustrate reductions in (a) habitat diversity, (b) species diversity and (c) genetic diversity.

Grade booster

Add in as much relevant detail as you can to your answer – know the figures for visitor numbers or the dates of legislation or the names of the key places in your case study.

Exam practice   answers: page 52

1 With reference to a specific case study, discuss the main causes for global reductions in biodiversity. (12 mins)
2 Critically evaluate the main management strategies that can be used to reduce declining biodiversity. (12 mins)

Answers
Biomes and ecosystems

Checkpoints

1 *Tropical broad-leaved forest* – temperatures 27–33°C all year round, precipitation 1500–2500 mm per year. *Temperate deciduous forest* – temperatures 0–25°C, with growing season of 8 months or more, precipitation 500–1500 mm per year. *Taiga* – temperatures –25–20°C with a growing season of 4 months or more, precipitation 300–1500 mm per year.

2 Maintaining biodiversity is important: to protect known resources, to protect resources we may not yet know about (e.g. future medicines), to stop unknown ecological changes, to maintain the diversity of species for humans to enjoy, and for moral reasons for the protection of species.

3 Main origins of East African tourists are European countries (especially UK, Germany and France), North American countries (especially USA and Canada), and Japan. The responsibility for erosion lies directly with the tourists who cause the damage, but indirectly with the tour companies who make a profit from tourists, and the national governments who exploit their resources for tourist income.

4 *Advantages* – preserve species, enable controlled breeding programmes, enable research, provide a visitor attraction and generate income.
Disadvantages – costly to establish, costly to maintain and protect, can be a target for poachers, may lead to overprotection and then to the need for culling, may increase erosion, and may change the natural ecosystem processes in the area.

Exam practice

1 Start by describing the scale and location of forest clearance, and outline the reasons for it in one paragraph. Then discuss the issues – group them under the headings of environmental, social, economic. Use examples, or a case study throughout. A good example or case study would be of forest clearance in Indonesia or in Brazil. Be sure to include the views of all the interested parties, e.g. LEDC government, logging companies, indigenous peoples, multinational companies, Western consumers, environmental groups, etc.

2 Show you understand the question by explaining briefly the growth and location, with reasons, for game parks. Then work through the issues (e.g. increasing animal numbers), describing the strategies used to deal with each, and their effects. Illustrate each with examples, or choose a particular case study that shows most of the key issues, e.g. game parks in Kenya or Zambia. Show the advantages and disadvantages of game parks in relation to social, economic and environmental issues.

Checkpoints

1 Living components of a pond: microscopic plants and animals (e.g. amoebae), insects (e.g. pond skaters), higher plants (e.g. pondweed), higher animals (e.g. fish, frogs), decomposers (e.g. bacteria), land-based animals (e.g. heron, water rat). Distinguish between living things (a) in the water, (b) in the bottom deposits. Non-living components: water, dissolved gases (e.g. oxygen, carbon dioxide), minerals (nutrients).
Inputs: streamflow, direct precipitation, minerals from erosion, temperature (energy).
Processes: living processes (e.g. movement, respiration, sensitivity, growth, reproduction, excretion, nutrition), chemical processes (e.g. solution), physical processes (e.g. temperature changes, transport of materials by water).
Outputs: water outflow, evaporation, living things (e.g. birds, seeds).

2 *Biomass* – the total mass of all living things or of a particular species in an ecosystem.
Net primary productivity – the amount of energy stored as plant material in an ecosystem.
Herbivores – animals that feed only off plants.
Carnivores – animals that feed only off animals.
Top carnivores – animals that feed off carnivores but are not preyed upon themselves.
Food chain – the route by which energy passes between living things by feeding.
Food web – the complex feeding inter-relationships between living things.

3 A tiger is a top carnivore. It requires a large territory (ecosystem) and biomass to support its energy needs.

4 Burning fossil fuels, quarrying limestone, clearing forest.

Exam practice

1 (a) Explain each part of the term. *Net* means it is the energy stored after life processes; *primary* that it is energy accumulated by plants. Overall it is a measure of how much energy is converted to biomass. Explain the units that NPP is measured in by saying what they mean, i.e. grams is biomass, m^2 relates it to a unit of area, and year is the measure of time over which it is measured.

(b) Choose examples of ecosystems to represent the list. Explain why each has a high or low NPP. Show the effect of temperature, water and nutrient availability, and explain the impact of each of these.

2 Describe *and* explain the carbon cycle. Use a diagram/flow chart. For each stage explain the processes, and then show how human activity interferes, e.g. how burning fossil fuels alters flows of carbon and adds it to the atmosphere. Explain what effect these changes to flows may have, e.g. the effect of increased carbon dioxide in the atmosphere and its impact on the 'enhanced greenhouse effect'. Don't forget to use real examples that you have studied, e.g. Amazon rainforest clearance.

Small-scale ecosystems

Checkpoints

1 Biotic elements are living things (e.g. plants, animals). Abiotic are non-living elements (e.g. water, minerals)
2 The pike is the top carnivore. The other fish will be carnivores or omnivores, but will be the food of the pike.
3 During the warm wet summer the population of birds will grow because there will be plentiful food, and a large proportion of the newborn birds will survive. If the winter is harsh, food will be in short supply, competition between the birds will be severe, and many of the birds will die, reducing the population.

Exam practice

1 Begin by clearly identifying the type of ecosystem you are using as your example (e.g. a pond, or disused railway line), then locate it by description and if possible by drawing a sketch map. Explain the principles of negative feedback and positive feedback, if possible using a diagram – negative feedback switches off change while positive feedback leads to permanent change. Finally, use examples from your chosen small-scale ecosystem to illustrate the ideas. For example, a new species of fish introduced into the pond will lead to competition with existing species. If the new species survives and replaces an existing species then this illustrates positive feedback. If the new species does not survive the competition then this illustrates negative feedback.
2 (a) *ecological niches* are a unique way of life and position within the food web of the ecosystem
   (b) *predator–prey relationships* – as the prey increase in number, so can the predator. If prey numbers decline, so do predator numbers
   (c) *competition* – competition occurs between species when a new species is introduced into an ecosystem, but mostly it occurs between individuals of the same species
   (d) *homeostasis* is the steady state where an ecosystem is not undergoing change.
Remember to use an example to illustrate each idea!

Ecological succession

Checkpoints

1 *Primary succession* – the development of an ecosystem from an initial state without plants and animals.
   Secondary succession – the development of an ecosystem from an existing ecosystem as a result of change in environmental conditions.
   Hydrosere – an ecosystem where water is the major controlling influence on its nature.
   Xerosere – an ecosystem where drought is the major influence on its characteristics.
   Psammosere – an ecosystem where the presence of sand is the major influence on its characteristics.

2 *Natural factors* – fire, drought, storm damage, flooding, disease.
   Human factors – clearance of vegetation, introduction of a new plant or animal species, pollution.
3 A plagioclimax community is one in which the main influence on its nature and characteristics is the interference of people.

Exam practice

1 (a) Describe the increase in species, quoting actual figures, and identifying when growth was fastest. Then use key words, e.g. *primary succession, pioneer species*, to show that you understand succession. Explain *why* species increase, and the impact of increasing numbers of plants and animals on food supply and predator–prey relationships.
   (b) (i) Start by explaining the examples you will use and locate these with sketch maps. Describe the process of dune succession. Name species at each stage, and draw a cross-section model.
   (ii) Identify the management issues in your chosen examples. List them to start with, then show how each is caused and managed. Distinguish between social, economic and environmental issues.
2 Identify key words in the quote, i.e. *management, communities*. The question wants you to show that you realise most British communities are plagioclimax. Choose examples and show how they are created and managed by people. The best examples will be ones where you can use fieldwork evidence and local knowledge. Discuss and explain secondary succession, what the climax community might be, and what the impact of people is in each of the changes. Try to show how the impact of human interference can be minimised, but also that such plagioclimax communities will only be sustained by management.

Threats to biodiversity

Checkpoints

1 *Species diversity* refers to the number of different species. *Genetic diversity* refers to the variations between individuals within a particular species.
2 The greater the biodiversity the greater the resource opportunities for people. There are then more possibilities for developing new foods, pharmaceuticals or industrial resources in the future, and the risk of major impact from plant diseases to existing food plants is reduced. There is also the aesthetic and spiritual value of maintaining a greater diversity of species and habitats.
3 Population growth and increasing standards of living and resource demands have used more land for agriculture, industry or housing or the clearance of ecosystems to provide resources such as timber – this leads to habitat destruction. Pollution, climate change, hunting and the introduction of 'exotic' species also impact upon diversity.

Exam practice

1 Describe the specific case study you are going to use, and explain its location and characteristics (for example an area of tropical forest in Malaysia). Show the range of processes affecting that habitat – for example, legal and illegal forest clearance for timber, clearance for agriculture (e.g. palm oil plantations), illegal hunting, pollution, climate change. Explain the main reasons for these changes e.g. economic need by the host country to earn income from selling timber and palm oil, or population growth.

2 Describe each of the main management strategies – biosphere reserves, zoos, gene banks, national and global agreements and legislation e.g. on trade of endangered species, and on climate change and pollution. Give a named example of where each strategy is being used. For each one, identify its main advantages and problems.
e.g. for biosphere reserves the advantages are protection of species and habitat, the possibilities of research, earning income from tourism. The problems are finding suitable locations, protecting locations from illegal exploitation, the costs of establishing and managing them, and the risk that it allows other areas to be destroyed or exploited.

Revision checklist
Biomes and ecosystems

By the end of this chapter you should be able to:

1	Describe and explain the characteristics of tropical forest and grassland ecosystems.	Confident	Not confident. **Revise** pages 40–41
2	Explain the main human impacts on tropical forest and grassland ecosystems and the main management strategies used to protect these biomes.	Confident	Not confident. **Revise** pages 40–41
3	Describe and explain the structural characteristics of ecosystems, including trophic levels and food chains.	Confident	Not confident. **Revise** pages 42–43
4	Describe the main processes in ecosystems, including energy flow and nutrient cycling.	Confident	Not confident. **Revise** pages 42–43
5	Explain the characteristics of small-scale ecosystems.	Confident	Not confident. **Revise** pages 44–45
6	Describe and explain the main processes of change in ecosystems, including: – negative and positive feedback – competition – predator–prey relationships.	Confident	Not confident. **Revise** pages 44–45
7	Explain and illustrate, using examples, the process of ecological succession, and the characteristics of seres, climax communities and plagioclimax communities.	Confident	Not confident. **Revise** pages 46–47
8	Explain the concept of biodiversity and the difference between habitat, species and genetic diversity.	Confident	Not confident. **Revise** pages 48–49
9	Describe and explain the main threats to biodiversity.	Confident	Not confident. **Revise** pages 48–49
10	Use examples to show the strategies that can be used to slow biodiversity decline.	Confident	Not confident. **Revise** pages 48–49

Environments

Geography at both AS and A2 requires you to study some of the major environments of Earth. This involves the study of the physical geography of these environments, including the processes that shape the landscape and the landforms that are produced as a result. It also involves the study of the human use of these environments, and the way that people organize their economy and society. Studying environments also shows the importance of studying people–environment issues, for the physical environment and its changes shape the way people use it, while people are changing the landscape through both local actions (for example in building new settlements) and global processes (for example through climate change). The main environments studied are those of coasts, tundra, polar, and desert.

Exam themes

- Coastal environments

- Threats to coasts

- Tundra and polar environments

- Glacial environments – glacial systems

- Glacial environments – glacial landforms

- Desert environments

Topic checklist

	EDEXCEL		AQA		OCR		WJEC	
	AS	A2	AS	A2	AS	A2	AS	A2
Coastal environments	O		O		O			●
Threats to coasts	O		O		O			●
Tundra and polar environments		●	O		O			●
Glacial environments – glacial systems		●	O		O			●
Glacial environments – glacial landforms		●	O		O			●
Desert environments		●	O		O			●

Coastal environments

Compared with other geomorphological environments, coasts change very rapidly. The processes involved here are very active – complete features such as mud banks and spits may form in little more than a century. Coastal erosion is removing cliffs and beaches at such a rate that local authorities are faced with significant management problems, individuals are faced with concerns about obtaining insurance for their property if it is located near the coast and coastal property values are falling.

Links

Threats to coasts, pages 56–57

Coastal processes

Waves are generated by the transfer of energy from the wind blowing across the sea surface. Their size and strength are related to:

→ wind speed
→ fetch (the distance over which they have travelled)
→ wind persistence.

Forced waves are driven ashore by the wind and are of two types:

→ constructive – flatter, longer periodicity (10 or more seconds between waves); these tend to move material up the beach
→ destructive – steeper, more frequent (3 to 6 seconds between waves); material is combed down the beach.

The jargon

Waves are described using the terms height, velocity, length and period. You should be able to use these terms confidently.

Action point

Draw a simple, labelled diagram to show the difference between the strength of the swash and backwash of constructive and destructive waves.

As waves approach the coast their direction is changed by friction with the sea floor to become nearly parallel to the shore. This is **refraction**. Material is transported along the beach by **longshore drift**. The movement of the wave up the beach is **swash** and the return movement down the beach is **backwash**.

Checkpoint 1

Write one or two sentences to explain each of these processes.

Coastal erosion processes

→ Wave quarrying or hydraulic action
→ Abrasion/corrasion
→ Attrition

(also **sub-aerial weathering**, **mass movement** and **human activity**).
The rates of erosion on a coast are influenced by the following:

→ rock lithology, structure and dip
→ local winds and water movement
→ the configuration of the coast leading to **concordant** or **discordant** coastlines.

The jargon

The lithology of the rock means its relative hardness, permeability and solubility.

Landforms of coastal erosion

The two main features produced by wave erosion are **cliffs** and **wave-cut platforms**. The cliff retreats as a result of undercutting and subaerial weathering, leaving behind a platform that is gradually lowered by corrasion. As the platform develops in a seaward direction the wave energy at the cliff face is reduced and consequently so is the rate of cliff erosion. The erosive power of waves is also concentrated on headlands of hard

Checkpoint 2

Explain the retreat of a headland and the landforms that you would expect to find at various stages in this process.

rock that retreat. The waves are able to attack the headland on three sides resulting in **geos**, **caves**, **blowholes**, **arches**, **stacks** and **stumps**.

Landforms of coastal deposition

Material is deposited along the coast when the load-carrying capacity of the sea is reduced, usually as a result of a reduction in wave energy and velocity. Many different landforms result.

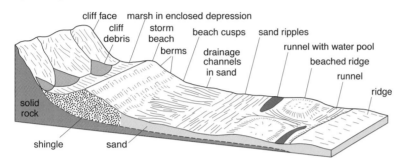

There are a number of other coastal depositional features:

→ spits and cuspate forelands
→ on-shore and off-shore bars and tombolos
→ salt marshes
→ sand dunes.

These features do not rely solely on the erosion of headlands for their supply of sediments. They are also supplied with:

→ material moved landwards from the offshore zone
→ river-borne sediments from inland
→ material from other beaches moved by longshore drift
→ beach nourishment.

Coastal changes

Many coastal landforms are largely the result of changes in the balance between erosion and deposition. This may be the result of a **change in sea level**, which may be either **isostatic** or **eustatic**. When sea level rises, the coastline is said to be **submerged** or **retreating** and lower parts of river valleys may be flooded to create **rias** or **fjords** and cliffs may be subject to renewed wave erosion. When sea level falls, **emerged** or **advancing** coastlines form, with landforms that were created at sea level but are now abandoned above sea level. These include **raised beaches**, **marine** or **raised platforms** and **abandoned cliffs**.

In addition to these changes along the coasts there are a number of threats to the coastal environment which are discussed on the next two pages.

Don't forget

The features of beaches that are shown here are all transitory because tides, winds and weather conditions are continually changing.

Action point

Draw up a table of each of these features of coastal deposition and the location of an example that you have studied.

The jargon

Isostatic changes are localised changes due to such things as post-glacial uplift of land. *Eustatic* changes are global in extent and affect all oceans equally.

Action point

Can you say whether each of the human influences on the beach environment would tend to cause more erosion or more deposition?

Exam practice   answers: page 66

1 Explain why beach features are transitory parts of the landscape. (5 mins)

2 Describe the process of longshore drift and one landform that is largely created by this process. (10 mins)

3 Use examples to explain the difference between concordant and discordant coastlines. (10 mins)

Threats to coasts

Action point

Make sure you know examples of each of at least two of the three threatened environments and can provide some information about the nature of the threat they are facing.

The coasts of the world are under threat. They are open systems in a state of dynamic equilibrium, which can easily be upset, particularly through human activity. They are some of the most densely populated parts of our planet and are experiencing great pressures from conflicting land uses such as leisure and tourism, industrial development, energy development, conservation, urban growth, ports and agriculture as well as some of the natural processes described in the last section. These areas are so popular because they can offer flat land with fertile soils, biodiversity and a favourable climate.

Pressure on coastal environments

The demand for coastal space is growing all the time and this is putting an increasing pressure on the natural environment. All coastal areas are susceptible to pressures, but the salt marsh, coral reef and mangrove swamp areas are particularly under threat. Salt marshes play an important role in providing animal and bird habitats and in reducing tidal energy. Coral reefs are found in shallow tropical seas and are living organisms. Mangrove swamps are rich ecosystems that are the breeding ground for fish and shellfish. They help to create new land by trapping sediments in the tree roots, offer protection against storm surges and are a source of fuel.

Coastal Environment	Threat
Salt Marsh	Reclamation for development, water pollution due to agricultural and leisure use, erosion by the "wash" from pleasure boats, pressure for new developments, e.g. marinas.
Coral Reefs	Quarrying for building materials, pollution, ocean acidification, tourism as they are highly sensitive to touch.
Mangrove Swamps	Clearance for timber, shrimp aquaculture and tourist developments.

The jargon

An EIA is designed to predict and then evaluate all the main effects of a new development on the environment.

The pressure on the coastal environment is at its most intense when the need to preserve the environment is matched against pressure for development. Each case has to be studied on its merits and the various economic benefits such as employment and local business income weighed against the destruction of or damage to coastal habitats. In the developed world it is always necessary to prepare an EIA (Environmental Impact Assessment) before a new development will be allowed to progress.

Managing coastal environments

Don't forget

Coastal management is made more difficult because there is no overall coastal planning authority and coastal processes take place across political boundaries.

Successful management of coasts involves achieving a balance between conservation and development, resolving conflicts between different users and, vitally, understanding and monitoring the sediment budget. The management of these areas is frequently about balancing socio-economic and environmental needs. Evaluating the costs and benefits is a complex task. Engineering schemes designed to protect one section of coast may have a detrimental effect further along the coastline.

The sediment budget

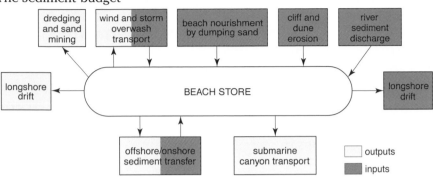

Action point

Make two lists of the inputs and outputs from the sediment budget diagram to show those that are influenced by human activity and those that are not.

How do coastal changes affect people?

→ Silt deposited in harbours and river mouths requires dredging.

→ Storm waves create damage to coastal settlements.

→ Longshore drift can remove tourist beaches.

→ Increased cliff erosion causes loss of land and properties.

→ Emerging coasts leave ports 'high and dry'.

→ Rising sea levels lead to more coastal flooding.

Checkpoint

Explain how the management strategies used in this diagram will affect the sediment budget for the area.

Engineering strategies

Hard engineering means the building of structures out of concrete, timber or rocks to control the force of the waves or the action of the tides and includes sea walls, revetments, Rip rap, gabions, groynes and barrages. These have several potential problems associated with them. They are expensive both to build and to maintain; may spoil the appearance of an attractive stretch of coast; defenses in one place can change the energy dissipation along the coast and lead to increased erosion elsewhere; sea levels are rising so fast that the defenses may not be adequate after a few years.

Soft engineering techniques appear more natural. They have been developed more recently as people have become more aware of and concerned about our environment. They include beach nourishment, dune regeneration, marsh creation, land use and activity management and managed retreat. They tend to be less expensive and are more aesthetically pleasing. However they also have some disadvantages such as the loss of land and property.

Action point

Make sure you are able to describe two contrasting coastal management schemes, one of which is based largely on hard engineering and another one based on soft engineering.

How do people's actions affect the coastal system?

People use the coastal zone for a variety of purposes and each of these affects the environment in a different way.

Human intervention along a section of coast

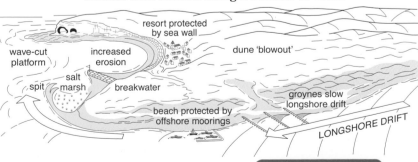

Examiner's secrets

It will be vital for you to know detailed case studies to enable you to answer questions on these topics.

Exam practice   answer: page 66

Describe the positive and negative features of a multi-purpose coastal management scheme that you have studied. (20 mins)

Tundra and polar environments

On the pole-ward side of the coniferous forests lie the tundra zones, which merge into the ice-covered polar regions. In the southern hemisphere the continent of Antarctica is a major wilderness region. Many regions have permanently frozen ground (permafrost) near the surface. These regions provide extreme physical challenges to people and raise many environmental issues. The main regions are:

→ the northern wilderness regions of northern Canada and Alaska, northern Scandinavia and northern Russia
→ Antarctica.

Action point

Find out the climate data for a number of places in the tundra or polar regions.

Tundra ecosystems

Tundra ecosystems are characterised by:

→ plants that can tolerate low temperatures, wet conditions and strong winds, such as grasses, mosses, dwarf willow and birch
→ plants that can reproduce in the short summer period
→ animals that can withstand the cold (e.g. arctic fox)
→ some areas of permafrost. This is permanently frozen ground. It develops in conditions of very cold temperatures (below –5°C) all year round, low precipitation totals so that snow cannot insulate the ground, and a limited vegetation cover.

Test yourself

Put the following species into a tundra food web diagram: arctic fox, caribou, lichen, reindeer, arctic wolf, polar bear, grass, arctic hare, moss, ptarmigan.

The northern wilderness regions

These regions are characterised by:

→ very low population densities and large unpopulated areas
→ some remnants of traditional cultures (e.g. the Inuit of Canada)
→ severe transport difficulties
→ harsh environmental conditions
→ an economy based on exploiting mineral resources and on some summer tourism (e.g. coastal Greenland)
→ severe social and economic deprivation amongst permanent residents.

Checkpoint 1

What are the main transport difficulties in northern wilderness regions?

Action point

Make some detailed notes on a case study of major mineral extraction, e.g. oil at Prudhoe Bay, coal in Spitzbergen, gas in Russia.

The principal environmental issues relate to mineral exploitation. Oil extracted at Prudhoe Bay, Alaska, for example, is transported by surface pipeline to Valdez in southern Alaska, raising problems of:

→ oil pollution from pipelines, especially in earthquake zones
→ oil pollution from tanker accidents (e.g. *Exxon Valdez*, in 1989)
→ the impact of oil installations and oil towns through pollution and waste disposal issues
→ disruption of animal migration patterns, as pipelines act as barriers, e.g. for caribou
→ damage to permafrost and plant communities by vehicles, or the use of heated pipelines for utilities or oil transport.

Antarctica

Antarctica is almost entirely covered by vast ice sheets, and supports only limited animal and plant life. It has no permanent inhabitants, and is managed by 39 nations under the terms of the 1959 Antarctic Treaty. Despite being a wilderness, Antarctica has many environmental issues, e.g.:

→ pressure to exploit its mineral resources; although exploitation has been prohibited for 50 years, the continent has vast resources of coal, and the continental shelf may have large oil reserves
→ the Antarctic 'ozone hole' resulting from pollution by MEDCs
→ the impact of tourism – cruise liners and visitors cause pollution and disrupt penguin and seal colonies
→ environmental damage at Antarctic bases – pollution from the US McMurdo base has been a problem, as has the construction of runways.

The table below compares the resources of the Arctic and Antarctic regions:

Resource	Arctic	Antarctic
Minerals	Oil (Canada), gas (Russia), coal (Spitzbergen), plus many minerals	Coal and many minerals
Fish	Cod, haddock and capelin – stocks under pressure	Only two species exploited
Whales	No commercial whaling	Some scientific whaling
Environment	Wilderness resource	Wilderness resource, valuable for scientific research

Strategies for managing Antarctica depend on the continuation of the agreements of the Antarctic Treaty.

> *"Any activity relating to mineral resources other than scientific research shall be prohibited for at least 50 years."*
>
> Antarctic Environment Protocol, 1991

Action point

For each of the issues described in the Antarctic environment, think through what might be done to address the problem.

Checkpoint 2

Why are countries so eager to have Antarctic bases and territories?

Checkpoint 3

What differences are there between the issues facing Antarctica and those facing the Arctic?

Grade booster

Group your ideas together, e.g. 'social factors' to show that you have a good overall understanding of what you are writing about. Think PEST – **p**hysical, **e**conomic, **s**ocial, **t**echnological factors – as categories to use.

Examiner's secrets

Including a simple sketch map to show the locations you are talking about is very helpful.

Exam practice   answers: page 66

1 With reference to a specific case study, discuss the social, economic and environmental issues that arise from mineral exploitation in tundra or polar regions. (12 mins)

2 'Wilderness areas should not be exploited for their economic resources.' Critically evaluate this statement and comment on its relevance to either northern Canada/Alaska or Antarctica. (15 mins)

Glacial environments – glacial systems

Today glaciers, ice sheets and ice caps cover about 10% of the land surface of the earth whereas during the Quaternary era this figure rose to 30%. Where areas are covered by ice, the land is being shaped by a number of unique processes.

The formation of ice

Checkpoint 1

Why will it probably take several hundred years for this process to take place in Antarctica or Greenland?

Glacial ice is formed as a result of the accumulation of layers of snow above the permanent snow line. Temperature changes lead to freezing followed by thawing and the snow becomes firn or névé. As the increased pressure of additional layers gradually squeezes out the air, the density of the material increases until after 20–40 years it becomes glacial ice.

Types of glacier

Glaciers are classified according to their size and shape.

→ Niche – small, in gullies on shaded hillsides.
→ Cirque – larger than niche, in armchair-shaped hollows in mountains; the glacier may spill out to feed valley glaciers.
→ Valley – larger masses of ice, moving downhill, usually following a former river course.
→ Piedmont – formed from the merging of valley glaciers in lowland areas.
→ Ice cap/ice sheet – enormous areas of ice spreading out from central domes; they may have nunataks projecting above them. Today these are found only in Antarctica and Greenland.

The jargon

A *glacier* is a slowly moving mass of ice that originated from an accumulation of snow.

The glacial system

Glaciers can be considered as systems with inputs, stores, transfers and outputs.

Action point

Make a table of the inputs, stores, transfers and outputs from a glacier system.

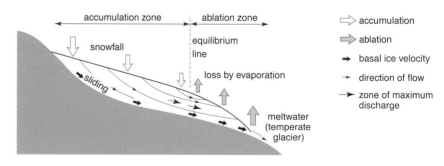

During a year the mass of the glacier will change considerably as shown in the diagram below. This is known as the glacial budget.

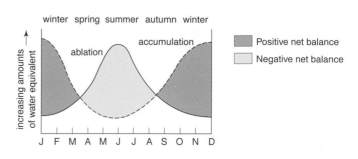

Glacial processes

Glaciers are extremely powerful agents of erosion. Erosion is most effective when the ice is moving and contains debris. Thus **warm glaciers** found in the temperate mountain areas are more effective than the **cold glaciers** of polar and subpolar regions.

Ice movement

Ice moves dramatically as **avalanches**, by **basal sliding** along the valley floor (important in warm glaciers) and by **internal plastic flow**. Movement is influenced by friction with the valley floor and sides, so the fastest movement is in the middle of the valley at the surface of the ice.

The processes of glacial erosion

→ Plucking – tearing away previously weathered pieces of bedrock.
→ Abrasion – polishing, scratching and gouging of the bedrock.
→ Frost shattering – breaking off pieces of bedrock by the alternating of freezing and thawing of water within joints in the rock.
→ Rotational movement – increased pressure resulting from pivoting of ice in a hollow.
→ Dilation/pressure release – upward expansion in rocks beneath the ice due to reduction in the overlying mass as rocks are eroded causes joints parallel to the surface.
→ Meltwater erosion – normal fluvial erosion below and beyond the ice.

The processes of glacial transportation

Anything from fine sediments to large boulders can be carried

→ supraglacially – on the surface of the glacier.
→ englacially – carried within the glacier or ice sheet.
→ subglacially – moved along at the base of the glacier.

The processes of glacial deposition

Drift is the name used to describe all material that is deposited under glacial conditions. There are two types:

→ **Till** is all material deposited directly from the ice. It is unsorted. It includes **boulder clay**, **erratics**, **moraines** and **drumlins**.
→ **Fluvioglacial deposits** are laid down by meltwater either within or beyond the ice and are sorted materials, often arranged in layers. These include **outwash sands** and **gravels**, **varves**, **kames** and **kame terraces**, **eskers**, **kettles** and **braided streams**.

Exam practice   answers: page 66–67

1 Describe how and explain why the mass of a glacier changes during the course of a year. (15 mins)

2 Explain how global warming will affect the mass balance of an alpine glacier. (10 mins)

3 Describe the characteristics and typical location of a drumlin. (10 mins)

Checkpoint 2

Can you explain why a warm glacier erodes more than a cold glacier?

Checkpoint 3

You should be able to describe how each of these three methods of ice movement takes place.

Action point

Can you explain why it is that these processes are likely to be most effective when temperatures frequently alternate above and below 0°C, rocks are well jointed, gradients are steep and where ice depths are great?

The jargon

Erosion is wearing away. *Transportation* is material being carried along. *Deposition* is material being deposited.

Grade booster

The media cover stories about glacier growth and retreat. Find a recent example to use in your answers.

Examiner's secrets

As with all A-level questions in geography the examiners are always going to credit your answer more highly if it includes appropriate examples.

Glacial environments – glacial landforms

Some of the most dramatic mountain scenery in the world is the result of the glacial processes described on the last two pages. Glaciers have also been responsible for shaping many of the lowland areas of the world and glaciated landscapes have had an important impact on human activity.

Grade booster

As with all geomorphological topics you are expected to be able to write about real places and not simply to know the theories of formation and location. You should be able to give named examples of features.

Landforms of glacial erosion

Many of the features of glaciated erosion are to be found in highland areas. The effect of weathering and erosion since the last ice age has altered the appearance of many of these landforms but this effect has been less significant where the last ice has disappeared relatively recently.

Nivation hollows are small hollows, usually containing a patch of snow. These become enlarged by the process of nivation. In locations where the climate allows the snow to remain throughout the summer these may develop into cirques.

In the highest areas there are **pyramidal peaks, cirques**, and **arêtes**. The most impressive landform is the **glacial trough** with its associated **hanging valleys**, **truncated spurs** and **rock steps**. When these features are dammed, **ribbon lakes** form and if they are flooded by a rising sea level a **fjord** results. Within the large glacial troughs smaller features such as **roche moutonnée**, **crag and tail** and **striations** are to be found.

During the main phase of glaciation the effects of erosion within the upland areas can be so immense that watersheds are breached by glaciers, with the result that original drainage patterns are altered and rivers may end up flowing in a different direction after the period of glaciation. This is **drainage diversion** or **river capture**.

Checkpoint 1

Can you explain the process of nivation?

Action point

Draw up a table containing names or locations of examples of all the features named on these two pages.

The formation of a cirque

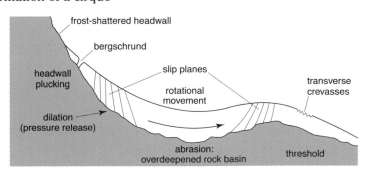

How glacial landforms can influence human activity

→ Provide routeways for communication in highland areas
→ Tourism income from visitors to spectacular scenery
→ Mineral extraction/quarrying
→ Dam sites for HEP
→ Frequently used for reservoir construction and water supply
→ Transhumance agriculture (now in decline in most areas)

Examiner's secrets

Many questions expect you to be able to discuss the effects that the physical landforms have on human activity.

Landforms of glacial deposition

Till deposits – unsorted material (unstratified)

→ Boulder clay
→ Lodgement till
→ Ablation till
→ Erratics
→ Moraine – lateral, terminal, push, medial
→ Drumlins

Fluvioglacial deposits – sorted material (stratified)

Beyond the snout of the glacier lies an area made up of sands, gravels and clays called an **outwash plain** or **sandur**, formed by deposition of material by meltwater streams during the summer or as the glacier melts. Other features include those shown in the diagram below and:

→ **varves** – annually deposited alternating layers of sediments
→ **kettles** – hollows or depressions in fluvioglacial deposits resulting from the melting of a block of ice that was trapped in the deposits
→ **braided streams** – choked due to seasonal discharge changes
→ **kames** and **kame terraces**
→ **eskers**.

(a) Glacial landscape

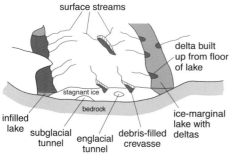

(b) Postglacial landscape

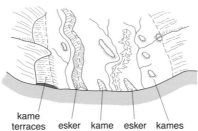

Links to human activity

→ Mineral extraction for building materials and subsequent landscape reclamation
→ Types of agriculture linked to variety of soil types
→ Location of settlements linked to landforms
→ Mountain wilderness areas and national parks – sustainability

Exam practice
answers: page 67

1 Describe the characteristics of a roche moutonnée and explain the processes that have led to its formation. (10 mins)

2 Explain how fluvioglacial deposits are different from glacial deposits. (5 mins)

Checkpoint 2

Describe how and explain why the characteristics of terminal, lateral and push moraines differ.

The jargon

Till deposits are materials dropped by the glacier. *Fluvioglacial deposits* are from water running out of the glacier.

Checkpoint 3

Use the diagram to help you explain the formation of eskers, kames and kame terraces.

Grade booster

Examination questions about glacial features often ask you to explain why the features are located where they are on the landscape. Use real examples to show your understanding of this.

Desert environments

Deserts cover 30% of the earth's land and occur in over 60 countries. Deserts, or arid zones, have less than 300 mm of rainfall per year, but semi-arid zones with 300–500 mm of rain per year have many similar management issues. The main issues in the world's hot desert regions are:

→ water supply and management
→ soil conservation
→ fuelwood management
→ desertification.

Action point

Check that you know the names and locations of the world's major deserts and semi-arid regions – then sketch their location on a world outline map as a test!

Hot desert ecosystems

Low precipitation and high temperatures support ecosystems where:

→ animals and plants have adapted to avoid the heat (e.g. nocturnal animals) and to find and retain water (e.g. cacti store water)
→ reproduction is timed to coincide with wet periods
→ plants can tolerate very dry and very salty conditions.

Checkpoint 1

What adaptations do plants and animals make to enable them to live in hot deserts?

Managing desert wildernesses

Water supply and management

Water supply problems arise because rainfall totals are low and often unreliable, and desert water is often saline. Stored water evaporates rapidly (30% of the water in Lake Nasser in Egypt evaporates each year). In addition, rapid population growth means demand for water increases rapidly, and traditional sources (e.g. groundwater wells) may be depleted or lost. Management approaches include:

→ rainfall harvesting, i.e. catching and storing as much rain as possible
→ increasing water extraction from groundwater
→ taking water from rivers, particularly those flowing into deserts from outside (e.g. the Aswan High Dam scheme on the River Nile)
→ desalinisation of sea water or saline lake or groundwater
→ piping water from 'water-rich' zones
→ maximising water use through trickle or point irrigation methods, and reducing evaporation from crops by mulching.

Action point

You will need to know a detailed case study of (a) a large multipurpose water management scheme (e.g. Aswan) and (b) a local-scale scheme. Make notes on one of each and learn them.

Soil conservation

Low and unreliable rainfall means that farmland may easily be damaged by salinisation, overgrazing or drying out. Desert soils are thin and lack much humus, so soil erosion by wind, or by rain when it occurs, is a major problem. Possible solutions include the maintenance of soil cover at all times by using cover crops or mulching, and managing irrigation very carefully. Tree planting may be helpful, so long as the species chosen are not highly demanding of water. Limiting numbers of pastoral animals to the land's carrying capacity is important.

Checkpoint 2

What are the main strategies to prevent soil erosion?

Fuelwood

The main source of energy in deserts in LEDCs is fuelwood. Limited availability and growing populations mean that trees and brush are cleared. This increases risks of erosion and desertification. International agencies are promoting the use of alternative fuel sources, e.g. solar power, biogas digesters, etc.

Desertification

Desertification affects the semi-arid regions of the world (e.g. the Sahel of Africa). It results from:

→ natural causes – unreliable rainfall and periods of drought cause marginal land to lose vegetation
→ population growth – increasing demand for resources means that unsuitable land is cultivated or land is overgrazed
→ poor land and water management
→ problems of fuel supply and the use of fuelwood

The figure below is a model of the desertification process:

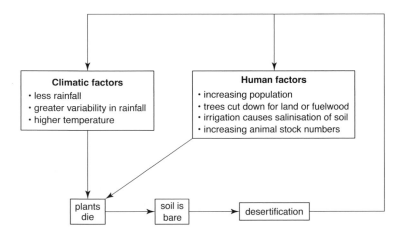

"The cost of desertification is around $26 billion a year."

Gaia Atlas of Planet Management, 1994

Checkpoint 3

What are the main alternatives to using wood for fuel in desert-margin countries?

Check the net

For case studies of the management of desertification, look at the UN Food and Agriculture Organisation web site: http://www.fao.org

Test yourself

When you have read these two pages, draw up a table listing the causes of each management issue covered, and the strategies for dealing with each one.

Exam practice   answers: page 67

1   How far are the main environmental management issues in deserts the result of low and unreliable rainfall? (15 mins)

2   Study the data below, which show the population and mean annual precipitation of Tanzania's semi-arid regions:

Year	Population (millions)	Average precipitation (mm)
1948	2.25	325
1958	2.61	360
1968	3.49	215
1978	5.98	240
1988	7.01	335
1998	7.98	340

(a) Describe and explain what the data show. (7 mins)

(b) With reference to a specific case study, critically evaluate the strategies that might be used to reduce the threat and impact of desertification. (15 mins)

Grade booster

Knowing two or three examples helps you to compare and contrast environments – learn case studies of both the Sahel and Australia as examples of water management in arid zones, and note the similarities and differences between them.

Examiner's secrets

The word 'critically' asks you to think of both the good and the bad features of what you are discussing.

Answers
Environments

Coastal environments

Checkpoints

1 Wave quarrying or hydraulic action is the energy of the breaking wave, abrasion/corrasion is the wearing away of cliffs by other material thrown by waves, attrition is the breaking down and smoothing of loose material.
2 As a headland retreats it might contain geos, caves, arches, stacks and stumps. You need to say how the processes in the previous Checkpoint act to erode the headland.

Exam practice

1 Transitory means short lived. Beaches are changed by every tide to some extent and as there are two tides each day they are truly transitory.
2 Longshore drift has been responsible, in part, for the formation of spits, tombolos and bars. The process is the movement of material along the beach by wave action. Material is moved up the beach at the angle of the swash and is then dragged down the beach by the backwash at right-angles to the beach line.
3 Concordant coastlines include the Dalmatian coastline of Croatia where there are long, thin islands parallel to the coast with narrow channels separating them from the mainland because the rock structure runs along the coastline. A discordant coastline is to be found in south-west Ireland. Here the structure of the rocks in the area runs across the line of ridges and valleys of the coast resulting in many inlets and long headlands. These two types of coastline are more likely to be created following a rise in sea level.

Threats to coasts

Checkpoint

Groynes and breakwaters will trap more sediments. New moorings will protect the beach so reducing sediment losses. There will be increased erosion at headlands as more sediment is trapped on the beaches, adding to the sediment budget.

Exam practice

Your answer should be divided into two clear sections explaining the positive and negative aspects and must focus on a coastal area that you know and have studied. Positive features might include reduced loss of beach material and benefits for tourism, reduced amounts of cliff erosion, a reduction in the need to dredge local rivers etc. The negative aspects might include loss of beach materials from further along the coast as a result of cliff protection measures, conflicts over the funding of aspects of the schemes, increased levels of erosion, changing ecosystems in the area with potential loss of employment for fishermen or other land users etc.

Tundra and polar environments

Checkpoints

1 The main transport problems are: huge distances; difficult terrain for road building, e.g. mountainous or permafrost areas; costs of maintenance; winter weather (grounding of aircraft, snow blocking roads and railways); cost of building transport systems.
2 Antarctica has huge potential resources. Countries want to make sure they have a 'say' in how these may be used in future and also wish to 'have their share' if resources can be exploited.
3 *Arctic* – main issues are political, since the territories of major superpowers converge there. Fewer political issues of resources, but many environmental issues of exploitation. *Antarctica* – political issues of ownership and future use of resources. Small-scale local environmental impact issues. Need for protection.

Exam practice

1 Start by describing the scale and location of mineral exploitation in tundra/polar regions, and the reasons for it in one paragraph. Then discuss the issues, grouping them together under the three headings of environmental, social, economic. Use examples, or a single case study throughout, and be sure to locate a case study with a sketch map. A good case study is the exploitation of oil from Prudhoe Bay in Alaska.
2 Start by showing that you know what wilderness areas are by trying to provide a definition with some different sorts of examples. In particular, discuss the meaning of the word 'wilderness' – is this wilderness for people to enjoy, or is it to be protected from people? Choose one of the regions and describe the arguments for/against mineral exploitation there. You could group your ideas as social, economic and environmental issues. Discuss the views of the range of interested parties, e.g. local indigenous people, multinational companies, environmental groups, etc. Finish by saying how far you agree/disagree with the statement, and why.

Glacial environments – glacial systems

Checkpoints

1 Ice formation requires alternating periods of freezing and thawing, which do not take place in Antarctica or Greenland.
2 Cold glacier – no melting, gentle gradients, temperatures lower than pressure melting point. Warm glaciers – melting so move faster due to lubricating water, especially in summer, steeper gradients, and base temperatures about same as pressure melting point.
3 Avalanches – sudden fall of ice rock and snow down a steep slope.
Basal sliding – movement of ice over the lubricated valley floor.

Internal plastic flow – plastic deformation of ice as individual particles move past each other under pressure and the influence of gravity.

Exam practice

1 Use the information on the two diagrams on page 60 to help you. You need to describe how the mass of the glacier changes (glacial budget) and relate this to the amount of precipitation and temperatures change between summer and winter, and then relate these to the processes of accumulation and ablation.

2 Global warming is raising the air temperatures so there will be more meltwater, greater evaporation, and more calving of ice sheets, especially where sea level is also rising, leading to a loss of mass. There may be more precipitation as weather patterns change, adding to the mass.

3 Drumlins are smooth elongated mounds of unsorted till, sometimes with a rock core. There is a steep stoss end facing the direction from which the ice came and a gently sloping down-stream lee side. Dimensions may be over 50 metres high and a few hundred metres long and wide. They are found singly or in swarms in glaciated valleys.

Glacial environments – glacial landforms

Checkpoints

1 Freeze–thaw action breaks up the rock under the snow and then meltwater removes the loose material as it flows out of the hollow.

2 Terminal – high ridge of unsorted material deposited by the glacier stretching across a valley marking the furthest point reached by the ice.
Lateral – an embankment of frost-shattered debris along the side of a valley carried along by the glacier from its point of origin.
Push – previously deposited moraine material that is shunted into a mound by a temporary ice advance. Some stones may be pushed into an upward position.

3 Eskers – long, narrow, winding ridges of sorted sands and gravel. Formed by deposition from subglacial streams.
Kames – undulating mounds of sand and gravel deposited unevenly along the front of an ice sheet; as the front retreats the mound collapses.
Kame terraces – flat areas of sand and gravel along the valley side. Deposited by streams flowing along the groove between the valley side and the glacier.

Exam practice

1 Roches moutonnées are solid pieces of bedrock that protrude from a glaciated valley floor. On one side they are smooth and gently sloping (stoss side) and on the other are rough, steep and uneven. As a glacier meets a resistant protrusion it slides over the stoss side polishing and scratching it. Then, by regelation (the refreezing of water into ice as the pressure is reduced on the downglacier side

of an obstacle), it plucks out rocks that have been loosened by freeze–thaw and pressure release from the lee side.

2 Fluvioglacial deposits are sorted into different sizes, are rounded as a result of being moved by running water and are usually laid down in layers. Directly deposited material is completely unsorted, containing materials of a variety of shapes and sizes and degrees of roundness.

Desert environments

Checkpoints

1 *Plants* – deep roots, small leaves with waxy layers, water storage (e.g. baobab), rapid reproduction cycle when water available, seeds withstand long periods of drought before growing. *Animals* – nocturnal or burrowing animals to avoid heat; can withstand long periods without water; sense organs keep out sand (e.g. camels' long eyelashes); shaped to avoid heat (e.g. camels' large feet).

2 Main strategies to avoid soil erosion are: maintaining soil cover at all times (e.g. by mulching, keeping soils wet), planting erosion controls (e.g. tree shelterbelts), keeping animal stock levels within the soil's carrying capacity.

3 Alternatives to fuelwood include: biogas digesters, mini-HEP schemes, fuelwood plantations, solar-powered cells, and animal power.

Exam practice

1 Show that you know the main environmental features of deserts with reference to precipitation, temperatures, arid and semi-arid regions, quoting actual figures. Locate them in relation to the main desert and semi-arid regions of the world. Then go through each of the main environmental management issues (e.g. desertification, water supply, soil erosion) and show what factors cause each. Be clear about the difference between natural and human factors in causing the problems. Use examples of each, or a case study to illustrate them all. A good case study is of desertification in the Sahel region of West Africa. Finish by drawing a conclusion about the question, i.e. answer the question.

2 (a) Describe the trends in each column and explain each one. Emphasise population growth and variability of rainfall. Talk about population growth curves (J-shaped and S-shaped curves). Discuss the rainfall reliability in semi-arid and desert margin regions. Stress that the figures show a potential issue, needing management to ensure that the problems that might result are minimised.

(b) Choose a case study – use a sketch map to locate it. Good case studies include the Sahel region of West Africa. Discuss each of the strategies to deal with desertification, grouping them as soil management, water management, agricultural management, population control, etc. Explain how each works and what social, economic or environmental issues each strategy may raise. For each, conclude why it is or is not successful (i.e. be critical!).

Revision checklist
Environments

By the end of this chapter you should be able to:

1	Describe and explain the physical and economic characteristics of coastal environments.	Confident	Not confident. **Revise** pages 54–55
2	Explain the main processes at work in coastal environments.	Confident	Not confident. **Revise** pages 54–55
3	Describe the main environmental challenges faced by people in coastal environments, and the main management strategies that can be used to reduce their impact.	Confident	Not confident. **Revise** pages 56–57
4	Describe the main processes in tundra and polar environments.	Confident	Not confident. **Revise** pages 58–59
5	Explain the main environmental issues in tundra and polar environments and the strategies available for managing them.	Confident	Not confident. **Revise** pages 58–59
6	Describe and explain the processes by which glaciers form and develop.	Confident	Not confident. **Revise** pages 60–61
7	Describe and explain the main landforms left by glacial processes.	Confident	Not confident. **Revise** page 62–63
8	Describe the location and climatic regimes of desert environments.	Confident	Not confident. **Revise** pages 64–65
9	Explain the main environmental challenges facing people in desert environments.	Confident	Not confident. **Revise** pages 64–65
10	Use examples to show the strategies that can be used to manage environmental issues in deserts.	Confident	Not confident. **Revise** pages 64–65

Natural hazards

Natural hazards are naturally occurring events or processes that have the potential to cause loss of life or property. They are not simply natural events, as without people they would form no kind of threat. It is the relationship between people and the environment that makes something into a hazard.

Many of the hazards experienced by people can be placed into three categories. **Tectonic** hazards are those related to plate movements and weaknesses. **Geomorphological** hazards are linked with atmospheric and tectonic processes. **Atmospheric** hazards are those related to short- or long-term variations in climatic patterns.

Exam themes

- The distribution of natural hazards
- The causes of natural hazards
- The impacts of natural hazards
- Managing impacts of natural hazards

Topic checklist

	Edexcel		AQA		OCR		WJEC	
	AS	A2	AS	A2	AS	A2	AS	A2
Principles and slope hazards				●		●	○	
Tectonic hazards	○	●		●		●	○	
Climatic hazards	○			●		●		●
Flood hazards and water conflicts	○			●		●	○	

Principles and slope hazards

Action point

Can you think of four hazards that fit into each of the three categories of this classification?

The jargon

Spatial describes distribution over an area. It may be on a large or small scale.

Action point

Use the diagram to describe the main events of a sudden impact hazard that you have studied.

Action point

Carry out research into the definitions of a disaster that are used by various organisations.

Classifying hazards

Why classify? Classifying the many different types of hazards allows people to focus on their key characteristics and in this way they can be more clearly understood. The human response can then be managed by governments, planners, hazard managers, insurance companies and any other people involved.

The simplest classification is into **natural** (such as a hurricane), **quasi-natural** (such as smog), or **human** (such as water pollution). A more commonly used classification is the one described by Bishop in 1998.

The five categories of this classification are:

→ **Type, or geophysical processes**
  → tectonic   → atmospheric
  → geomorphological → biological
→ **Cause** – natural, quasi-natural or human (see above).
→ **Magnitude and frequency** – the scale of the events and how often they occur. Low-magnitude events are more common than high-magnitude events. Some hazards, e.g. hurricanes, are **seasonal**, whilst lightning and fires are **random**.
→ **Duration of impact and warning time** – there may be a **sudden impact** or a **slow (creeping) onset**, or somewhere in between. It is also important to consider the longer time scale of impacts on human systems as shown in the diagram below.
→ **Spatial distribution** – some hazards only affect certain areas, e.g. hurricanes or tectonic hazards, whereas others are more widespread, e.g. river flooding. There is also a great variation in scale from local to international.

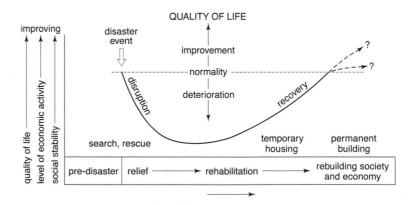

Hazard or disaster?

Some hazardous events take place in areas where the population is not vulnerable so they are not disastrous. In other areas people's lives are adapted to cope with natural processes when they are operating normally, but when events become extreme they may develop into a disaster. The difficulty is in defining a disaster.

→ Monetary values make comparison difficult over long periods.
→ Using numbers of deaths and injuries is a subjective approach.
→ The amount of damage varies greatly depending on the levels of technology and development in the country affected.

Mass movement or mass wasting

This is the movement of material downslope due to gravity, which is often almost imperceptibly slow. Water is usually present in the regolith and lubricates the particles of weathered material. Classifications of flows may be by speed, i.e. fast to slow, or by type of movement, i.e. flow, slide or heave. The most accepted classification of mass movements is that of Carson and Kirkby (1972) shown below.

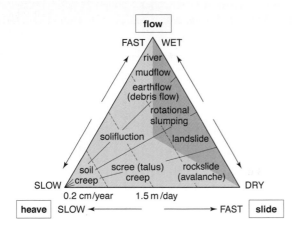

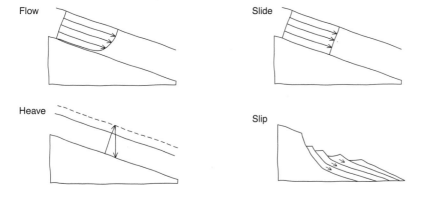

Action point

Construct a table of the types of mass movement shown in the top diagram and give a brief explanation of each. Make sure you think about the ways that each movement is different from others.

Checkpoint 1

With reference to the bottom 4 diagrams, explain the difference between flow, slide, heave and rotational slips.

Checkpoint 2

Can you explain why people in LEDCs are not always more vulnerable to natural disasters than people in MEDCs?

Factors influencing mass movement

→ Internal strength of regolith
→ Volume of loose material present on the slope
→ Slope angle
→ Water content of material/rainfall amounts
→ Human activity
→ Earthquakes.

Watch out!

Some of the examination syllabuses include human hazards such as disease and crime, but most stick to natural hazards!

Exam practice   answers: page 78

1 Explain how perception of a natural hazard may vary according to the level of economic development in the country affected. (10 mins)
2 Explain the difference between a hazard and a disaster, using the example of river floods. (10 mins)

Examiner's secrets

Use up-to-date examples. Cyclone Nargis and the Sichuan earthquake in China are examples from 2008. Keep up to date by using websites such as BBC News.

Tectonic hazards

Action point

You must be able to describe and explain the location of all these tectonic hazards. You should also be able to explain that they do not only occur at plate boundaries, but also in other locations, and that some are quasi-natural.

Tectonic hazards are those related to plate movements and weaknesses. These hazards create substantial risk to people, their possessions and buildings. The occurrences of these hazards are closely linked with the location of tectonic events, which vary in frequency and severity, and with the vulnerability of the population concerned.

The major tectonic hazards

Earthquakes

You need to be able to use the terms **focus**, **thrust fault** and **epicentre** correctly and to be able to explain the main hazards of **ground movement**, and also the secondary hazards of **soil liquefaction**, **landslides**, **tsunamis** (see below) and **avalanches**.

Each earthquake is different in the way that it is hazardous to humans and human activity. You should learn the details of the damage and consequences of your case studies in as much detail as possible. Management strategies include earthquake-resistant engineering, preparation within the community, land use planning, insurance and aid. Currently it is impossible to predict the occurrence of earthquakes accurately, but as technology and research improves this is being achieved with greater accuracy.

Earthquake hazards can be either **primary**; a direct result of the ground movement, such as building collapse, casualties or service pipeline rupture or **secondary**; effects which are subsequent to the actual movement such as diseases, tsunamis or infrastructure problems.

Checkpoint 1

What is the difference between the Mercalli and Richter scales?

Tsunamis

These are enormous waves that have been created by a large disturbance of sea water. These can be caused by:

→ an earthquake involving vertical movement of the seafloor
→ volcanic explosions
→ large landfalls.

They are most dangerous when they are generated close to a land area. The main effects are:

→ hydrostatic effects – objects and structures are carried by the waves
→ hydrodynamic effects – objects are torn apart and washed away
→ shock effects – battering by material carried in the water.

Checkpoint 2

Where would you expect to find most tsunamis and why?

Checkpoint 3

Why do you think that tsunamis are more dangerous when they are locally generated?

People are able to modify their vulnerability by predictions and warnings based on earthquake activity, careful planning of land use in vulnerable coastal areas, and building tsunami-resistant buildings. Following the 2004 Boxing Day tsunami in South East Asia, a series of tsunami warning systems has been established worldwide. These allow information about seismic activity, which is likely to cause tsunamis, to be transmitted rapidly to areas which are likely to be affected. This in turn will allow people to move inland or take shelter in appropriate buildings.

Volcanoes

The hazard potential of a volcano depends upon the way that it erupts and the type and amount of material that is released. The various hazards include **lava flows**, **pyroclastic flows**, **ash** and **tephra fall**, **gases**, **lahars**, **mudslides** and **volcanic landslides**. Volcanoes can affect all people in an area irrespective of their social or economic differences. Human responses are usually limited to monitoring, warning and evacuation. However, there have been successful attempts to divert lava by explosions and to cool it down by spraying it with sea water.

Impact and management of tectonic hazards

Tectonic hazards vary in their impact and in the way that they are managed.

Demographic, economic and social impacts of tectonic hazards

The scale of the **impact** of any tectonic hazard depends on the following characteristics of the area that is affected:

→ Level of economic development
→ Level of technological development
→ Population density
→ Levels of preparedness
→ Education and training about the risks involved

People's **socioeconomic circumstances** can cause them to be more vulnerable to the risks of tectonic hazards, for example, if they live in a very densely populated area, if they are poorly governed or if they are poor.

The scale of impact of tectonic hazards

Tectonic hazards have an impact at a variety of scales. If they affect a rural area this often means that less damage and loss of life occurs than if they affect an urban area. Some of the effects can be very **localised**, such as a lava flow on the slopes of mount Etna in Sicily, whilst others can affect whole regions, such as large earthquakes in Less Economically Developed areas (e.g. Kashmir earthquake in 2005).

How do people react to tectonic hazards?

People's reaction to tectonic hazards can be divided into short term and long term. Short term reactions are those that take place in the first few hours after a tectonic event. These are largely emergency rescue procedures. In areas that are prepared for such events, these activities are well planned and largely efficiently organized. Long term responses include managing the repairs and reconstruction, insurance and compensation payments and medical needs, as well as planning for the ability to deal with any repeated event through the training of the emergency services and the public in general, and the control of building regulations. In poorly prepared or more vulnerable areas the management of the after effects of a tectonic event is more likely to result in a **disaster**. The ability of the people of an area to manage these situations often depends upon their level of ignorance, degree of choice and most importantly their level of economic development.

Exam practice   answers: page 78

1 Describe the causes, consequences and management strategies associated with a tectonic hazard that you have studied. (20 mins)
2 Explain why tectonic events do not always result in disasters (20 mins)

Example

Study the Mount Pinatubo eruption. It is an excellent example of a large volcanic eruption and shows how prediction and response can be an effective management of a hazard.

Action point

Research the hazards that are associated with a jökulhlaup, which is an Icelandic word for a sudden outburst of meltwater caused by ice melt as a result of volcanic activity.

Action point

You must be able to describe a variety of management strategies that are used to reduce the impacts of tectonic hazards and comment on their likely degree of effectiveness.

Grade booster

Examiners will be hoping to read answers which clearly demonstrate detailed knowledge of how people have manipulated and how they have responded to events. You will need to know specific facts about particular events.

Examiner's secrets

For all these hazards you should be able to describe examples from an MEDC and an LEDC.

Climatic hazards

A **hazard** is a natural event that threatens both life and property – a **disaster** occurs when the hazard takes place and human life and property are put at risk. Weather hazards are very varied. They can be natural or man-made, local or global.

Checkpoint 1

Many people choose to live in hazardous areas. Why?

Types of weather hazard

→ Fog
→ Snow and ice
→ Droughts
→ Floods
→ Frosts
→ Hail
→ Hurricanes
→ Tornadoes

It is possible to characterise weather hazards and disasters in a number of ways:

1. **Magnitude** – the size of the event.
2. **Frequency** – how often an event of a certain size occurs.
3. **Duration** – the length of time that an environmental hazard exists.
4. **Areal extent** – the size of the area covered by the hazard.
5. **Regularity** – some hazards are regular, such as cyclones, whereas others are much more random, such as tornadoes.

Fog

Checkpoint 2

Why do urban areas experience a high incidence of fog?

Fog persists longer when there is a temperature inversion – i.e. when cold air at the surface is overlain by warm air. This is common in high pressure conditions, in valleys and over urban areas. Cold air, being denser, is unable to rise – thus the fog persists. In areas where there are large concentrations of smoke, sulphur dioxide and other pollutants, smog is formed.

Fog is a major environmental hazard – airports may be closed for many days and road transport is hazardous and slow. Freezing fog is particularly problematic. Large economic losses result from fog, but the ability to do anything about it is limited. This is because it would require too much energy (and hence cost) to warm up the air or to dry out the air to prevent condensation.

Tropical cyclones: hurricanes

The jargon

A *hurricane* is a tropical low pressure system. A *cyclone* is a mid-latitude depression, which is also a low pressure system.

Hurricanes are intense hazards that bring heavy rainfall, strong winds and high waves, and cause other hazards such as flooding and mudslides. Hurricanes are also characterised by enormous quantities of water. This is due to their origin over moist tropical seas. High-intensity rainfall, as well as large totals (up to 500 mm in 24 hours), invariably cause flooding. Their path is erratic, so it is not usually possible to give more than twelve hours' notice. This is insufficient for proper evacuation measures.

For hurricanes to form a number of conditions are needed:

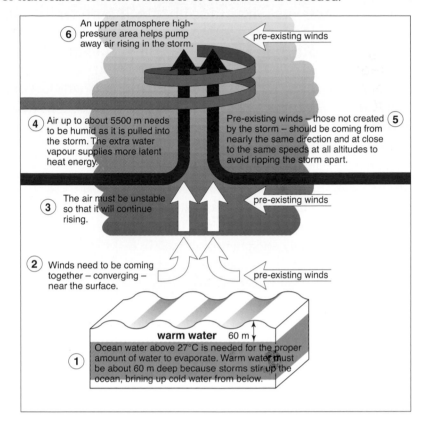

6 An upper atmosphere high-pressure area helps pump away air rising in the storm.

pre-existing winds

4 Air up to about 5500 m needs to be humid as it is pulled into the storm. The extra water vapour supplies more latent heat energy.

5 Pre-existing winds – those not created by the storm – should be coming from nearly the same direction and at close to the same speeds at all altitudes to avoid ripping the storm apart.

3 The air must be unstable so that it will continue rising.

pre-existing winds

2 Winds need to be coming together – converging – near the surface.

pre-existing winds

warm water 60 m

1 Ocean water above 27°C is needed for the proper amount of water to evaporate. Warm water must be about 60 m deep because storms stir up the ocean, brining up cold water from below.

Check the net

Visit the US National Hurricane Center at: www.nhc.noaa.gov

Hurricanes create a major threat to human life, property and economic activities. They are a seasonal hazard, peaking between June and November in the northern hemisphere. Because of their impact, and the cost of their destruction, they are monitored intensely by satellite, and hurricane paths are predicted by complex computer programs.

Tornadoes

For a tornado to occur in the USA, a number of factors need to occur simultaneously. These include:

→ a northerly flow of marine tropical air (from the Gulf of Mexico) which is humid and has temperatures at the ground in excess of 24°C
→ a cold, dry air mass moving down from Canada or out from the Rocky Mountains at speeds in excess of 80 km/h
→ jet stream winds racing east at speeds in excess of 390 km/h.

These three air masses, all moving in different directions, set up shearing conditions, imparting spin to a thundercloud.

Checkpoint 3

How do tornadoes and hurricanes differ?

Grade booster

You do not need to refer to the latest hazard to score most marks. You need examples that illustrate the many different aspects and impacts of a hazard. Some of these examples may be a few years old although some may be relatively recent such as Cyclone Nargis and Hurricane Katrina.

Exam practice
answers: pages 78–79

1 Why is fog an environmental hazard? (5 mins)
2 (a) How are hurricanes formed?
  (b) Why are hurricanes considered to be hazards? (15 mins)

Flood hazards and water conflicts

A river floods when its discharge exceeds its capacity. Every river and its drainage basin have their own characteristics, some of which are more likely to cause flooding than others. It is only when one or more of these features combine in the same basin that floods are likely to occur. Devastating floods are not limited to LEDCs, as the Boscastle flood in the UK in 2004 indicates. As demand for water increases, there are growing conflicts over this valuable resource.

Causes and impacts of flooding

Human and physical characteristics of drainage basins can lead to an increased risk of flooding. Physical factors include **excessive levels of rainfall** over a prolonged period of time, **intensive precipitation** over a short period of time, **snow melt** and **severe storms** associated with low pressure systems. Human activity in drainage basins can increase the likelihood of flooding by **urbanisation**, **deforestation** and the **engineering** that has been done to manage river flow such as the building of embankments, dams, and the channelisation of rivers. In recent years **global warming** has been blamed for climatic changes which have led to more rainfall and more intense rainfall events.

Areas that are at risk from the flood hazard exist in both MEDCs and LEDCs, although the ability to cope with and manage flooding events is usually more advanced in MEDCs.

Checkpoint 1

Explain how urbanisation can lead to greater likelihood of flooding.

Managing the flood hazard

The flooding hazard can be managed by warning systems based upon prediction and forecasting, preparation and training of local people, and by engineering solutions.

Hard engineering schemes	Soft engineering strategies
Dams	Changes in land use management on flood plains
Straightening	Wetland and river bank conservation
The building up of levees	River restoration
The creation of diversion spillways	Upland reforestation
Flood storage basins	Improved forecasts and warnings systems

Water conflicts

People need water to survive, but not everybody has enough and some people have access to more than they need. There are physical factors which dictate how much water is available in an area and the way that these resources are managed or mis-managed by people also affects potential supplies. 47% of the land area of the world (excluding Antarctica) is to be found in drainage basins that cover parts of more than one nation. Water supply can come from surface or groundwater supplies and is largely controlled by the physical factors of **geology**, **climate** and **river systems.** However, as populations rise and economies develop, the demand for water increases, a demand which cannot always be met from local supplies. This is leading to particular supply problems and to conflicts concerning those who manage the supplies:

→ in rapidly growing cities like Mumbai, which is located on a peninsular requiring enormous investment in a supply network
→ in whole regions of countries like India and China where economic development is creating greater domestic, agricultural and industrial demands for water
→ water supplies can become polluted
→ over abstraction can lead to salt water incursion affecting agriculture
→ extravagant actions of some wealthy people can lead to a lack of supply for those less well off.

Insecure water future

As the deficit of fresh water supplies grows there is likely to be an increasing conflict between various players or actors in the supply chain. These include **water companies**, **environmentalists**, **individual citizens**, **governments** and **manufacturing industries**. As the demand for water increases there are a number of different approaches to meeting the changing demand for fresh water. These solutions include **diverting supplies**, **increasing storage**, **water conservation**, **recycling water**, **restoration of lost supplies** and the use of **better technology** in water transfer or desalination. All of these options have costs and benefits of an economic, cultural and environmental nature.

International issues

In any area where the demand for water exceeds the supply there is the potential for conflict. In some parts of the world this is causing tension between nations such as in the Jordan river basin in the Middle East, the waters of which are relied upon by Israel, Jordan and Palestine, but are controlled by Israel. The river Brahmaputra is being argued over by India and China as the Chinese wish to alter its flow to suit their needs before it flows into India where it is a vital source of fresh water. Water supply is becoming an increasingly important political issue as an increasing number of treaties are being signed to protect and develop supplies. Fresh water is frequently being moved across borders, either from one region of a country to another, or between nations.

Grade booster

The best answers about water conflicts will demonstrate empathy for the different people involved in the management of water supplies. Try to think about the issue as if you were the people involved. How would you react and why?

Checkpoint 2

How have Majorca and Gibraltar overcome their shortages of fresh water?

Action point

Research an example of international water movement.

Exam practice answers: page 79

1 Explain how a river flood might have economic, social and environmental impacts in the affected area. (20 mins)
2 As the demand for water increases, explain the possible conflicts that might arise. (15 mins)

Examiner's secrets

To be able to answer all examination questions about flooding you should have two contrasting case studies prepared, one from an LEDC and one from and MEDC.

Answers
Natural hazards

Checkpoints

1 Flow is a slow, almost continuous movement downslope of material retaining moisture. A slide is a rapid movement of material along a well-defined plane such as a landslide. Heave is the raising of particles at right-angles to the slope and the subsequent falling down the slope due to the effects of gravity. Rotational slips are a combination of slides and flows.

2 Many people in LEDCs live in simple buildings that collapse around them without causing too much loss of life whereas people in MEDCs use concrete and bricks that cause far more injuries when they collapse. People in LEDCs may not be as vulnerable to financial loss as they own little compared with people in MEDCs, however, their livelihood may be completely destroyed.

Exam practice

1 Perceptions of a natural hazard may vary between acceptance, domination and adaptation. In LEDCs there is a lower ability to understand the hazard in detail and a lower ability to undertake technological solutions and therefore they are likely to view a hazard with acceptance or maybe adaptation. Depending upon the individuals or the organisation concerned, people in MEDCs are more able to understand and control hazards and they are likely to view hazards with either domination or control.

2 Living on a fertile floodplain with regular small floods is hazardous because the people know that a flood of larger, damaging proportions is possible. They benefit from the fertility and irrigation water that the river floods bring. The disastrous flood may bring loss of life, crops, property and communication problems.

Tectonic hazards

Checkpoints

1 The Mercalli scale is a descriptive scale used to describe the intensity of the shaking of the ground during an earthquake. The Richter scale is logarithmic and measures magnitude or total energy release of an earthquake.

2 Tsunamis are usually found along coasts around subduction plate boundaries as it is in these locations that the sea bed experiences vertical movement that causes most of these events.

3 This is because they move very quickly and there is little time for people to evacuate. Some 99% of deaths resulting from this hazard are caused by locally generated tsunamis. The greater the distance travelled, the longer time there will be to warn and evacuate people.

Exam practice

1 For questions like this it is essential that you know detailed case studies. Here you are asked to describe the causes. You need to refer to tectonic activity/ plate movements that were involved in the event. The consequences are the effects the event had on human and natural environment. Include longer-term effects such as loss of employment opportunities or the impact of insurance claims. When discussing the management strategies ensure that you include modifying the event, modifying vulnerability and modifying the loss.

2 To answer this question you need to explain that disasters are the result of a hazard becoming reality as an event occurs which causes deaths and damage to property and the environment on a significant scale. Tectonic events can occur in very remote areas either on land or at sea and not cause significant damage or deaths and so would not be a disaster. Volcanoes can erupt in a slow, non-explosive manner, as Etna and Muana Loa usually do, and these events may well not be disasters if little or no property is damaged. Large earthquakes can easily become disasters in areas where there is a weak economy, lack of preparedness, poorly constructed buildings and greater likelihood of hunger and disease as is found in many LEDCs. These situations are not usually common in MEDCs and so similar tectonic events do not become disasters.

Climatic hazards

Checkpoints

1 These areas have other benefits that outweigh the hazard problem, e.g. San Francisco is worth living in even with fog!

2 Cities produce a lot of heat and so warm air rises above them. On a cold night cold air is trapped beneath this layer. Pollution particles also encourage condensation.

3 A *tornado* is an intense low pressure system found over land. It develops as a result of the contrast between very cold air and very warm air. In addition, jet stream activity causes an intense spiralling wind system with winds of over 500 km/h. By contrast, *hurricanes* can develop as tropical low pressure systems over oceans. They are driven by warm sea waters and the spin of the Coriolis force. When they hit land they are cut off from their supply of heat (the warm waters) and so begin to die out, having first caused heavy rainfall, strong winds, flooding and mudslides in the land areas in their path.

Exam practice

1 Fog can cause large amounts of pollutants, such as SO_2 and particulates, to remain in the lower atmosphere, thereby affecting people's health. It makes driving dangerous – and aeroplanes and ferries may have to be cancelled. There is economic loss as well as a potential impact on health.

2 (a) Hurricanes develop over warm tropical seas. Intense heating causes large-scale evaporation of water. Once the low pressure system develops it is fuelled by evaporation. Winds blow inwards to the low pressure area. The whole system moves westwards. Once the hurricane reaches land it begins to die. This

is because it no longer receives any latent heat from condensation of water.

(b) Hurricanes are hazardous on account of:
- high winds with gusts of over 120 km/h
- heavy rain (up to 500 mm in 24 hours)
- a rise in sea level (10 cm for every drop of 10 mb)
- storm surges.

Flood hazards and water conflicts

Checkpoints

1 Urbanisation leads to a greater risk of flooding as large-scale paving and construction of buildings have taken place. This is designed to get rid of water as soon as possible, resulting in large amounts of water entering the local rivers very quickly, resulting in a greater than usual discharge and thus the likelihood of rivers bursting their banks. The concreting over of large areas also reduces the opportunity for water to infiltrate and travel more slowly through the soil and rocks into the river, again resulting in a more rapid and sudden discharge of water into the local drainage network.

2 Majorca has seen its demand for water increase rapidly as the tourist industry on the island has grown. There is such a deficit now that supplies are brought to the island from mainland Spain in tanker ships. Gibraltar has solved its water supply problems by creating large reservoirs to store rainwater inside the famous "Rock" of Gibraltar.

Exam practice

1 River floods may have a series of impacts. To answer this question you need to know detailed case studies and be able to describe economic impacts such as loss of farmland and livestock leading to loss of agricultural income, increased insurance policies or even the difficulty of obtaining insurance at all. Damage to property and businesses in urban areas followed by employment issues can also be serious impacts. Socially people's lives can be disrupted in many ways, such as having to leave their home, sometimes for many weeks after the flood, loss of employment or even bereavement. Environmental impacts include the disruption to the wildlife in the area, vegetation damage and the need to clean up after such events in urban areas.

2 The increasing demand for water is leading to conflicts such as industry, agriculture and domestic users competing for a dwindling resource in one area. International conflicts arise as rivers which provide supplies to one nation may be flowing through another, and the countries may wish to use the water for non–complementary uses, or even in too large a quantity for any to remain in the river further downstream. Increasing supplies by building more reservoirs creates conflicts between the users of the land which is to be flooded and the supply agencies. Desalination is an option, but conflicts arise due to the heavy energy requirements of such schemes. As supplies dwindle, water becomes a more precious resource and the ability to pay for water will be an increasing issue for poor people.

Revision checklist
Natural hazards

By the end of this chapter you should be able to:

1	Describe the main types of hazards and their main characteristics.	Confident	Not confident **Revise** page 70
2	Explain the main hazards related to slopes and slope processes.	Confident	Not confident **Revise** page 72
3	Describe and explain the main hazards related to plate tectonics.	Confident	Not confident **Revise** page 72–73
4	Describe and explain the main climatic hazards.	Confident	Not confident **Revise** page 74–75
5	Illustrate the hazards related to floods and water conflicts.	Confident	Not confident **Revise** page 76–77

Pollution

Pollution occurs when there is too much unwanted material in a specific place at a specific time. It can take many forms – such as air-, soil- and water-pollution. It can have a damaging effect on human health and on livelihoods. The type and amount of pollution varies with the type of country (MEDC/LEDC/NIC) and the level of industrialisation. For example, pollution from agriculture, such as eutrophication and soil erosion, is very different from pollution from industry, such as acid rain. As a country develops, the amount of pollution tends to increase. There are ways of tackling pollution but these can be quite costly. In addition, many MEDCs may out-source their industries to countries where legislation concerning pollution is lax. It is an increasing problem.

Exam themes

- Types of pollution
- The causes of pollution
- Pollution and economic development
- The effects of pollution
- Ways of managing pollution

Topic checklist

	Edexcel		AQA		OCR		WJEC	
	AS	A2	**AS**	**A2**	AS	A2	AS	A2
Principles of pollution		●		●		●		●
Water pollution		●		●		●		●
Atmospheric pollution: acid rain		●		●		●		●

Principles of pollution

Pollution is the contamination of the earth/atmosphere such that normal environmental processes are adversely affected. Pollution can be natural, such as from volcanic eruptions, as well as human in origin. It can be deliberate or accidental.

What is pollution?

The levels that constitute 'pollution' can vary. For example, decomposition is much slower in cold environments and so oil slicks pose a greater threat in Arctic areas than in tropical regions. Similarly, levels of air quality that do not threaten healthy adults may affect young children, the elderly or asthmatics.

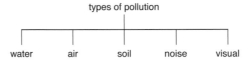

types of pollution: water, air, soil, noise, visual

Checkpoint 1

What are the natural sources of pollution?

Pollution leads to:
→ death
→ declining water resources
→ poor air quality
→ decreased levels of health
→ reduced soil quality
→ contamination of ecosystems.

Action point

Make a list of the social, economic and environmental costs of pollution, with examples.

Waste has increased in all European countries. Major sources of waste include agricultural, industrial, municipal and mining activities. The composition of waste is also changing and increasingly includes plastics and packaging materials. In Europe, most waste is disposed of in **landfills**. Without proper management, these can release pollutants into the soil and groundwater. In addition, carbon dioxide, methane and other toxic gases may be produced in landfills.

Check the net

Visit the BBC's State of the Planet at: www.bbc.co.uk/nature/programmes/tv/state_planet/pollution.shtml

Managing pollution

It is difficult to develop any form of pollution control when it is easy to estimate the **cost** of controls, but more difficult to assess (in monetary terms) the **benefits** of protecting the environment. In addition, it is difficult to assess the actual costs of pollution and to decide who should bear the costs.

It is even more difficult to develop **cross-frontier strategies** when dealing with pollution. The '**ecological time lag**' means that pollution problems are often not recognised until it is too late to do anything about them, let alone decide on a course of remedial action. Unless **point sources** can be targeted it may be impossible to treat pollution effectively. There is no point in treating symptoms, such as acidified lakes with lime, if the cause – the emission of acid materials – is not tackled.

The increase in vehicle exhausts and in sewage waste are notable failures to tackle the pollution problem. By contrast, the reduction of detergent phosphates and the decline of CFCs are good examples of successes.

Most synthetic cleaning products come from petrochemicals. Many domestic cleaners contain bleaches and perfumes. Early detergents did not break down rapidly in the environment, but built up in streams and sewage plants (foaming at the surface is characteristic). The production of biodegradable detergents in the mid-1960s reduced the problems, but an additional problem was the use of phosphates to soften hard water and to reduce its acidity. The phosphate-rich waste water accumulated in surface waters and groundwater, leading to eutrophication of streams and lakes. The only way of avoiding this was the development of phosphate-free detergents.

Pollution and economic development

There are a number of views and issues concerning the link between economic and social development and pollution:

→ Is pollution a necessary effect of growth?
→ Is it the price of progress?
→ Are economic development and environmental management two opposing themes?
→ Are they merely a battle between short-term profits and long-term costs?

Although pollution is associated with **capitalist development**, it is not restricted to capitalist countries. The communist countries of the former Eastern bloc have large-scale pollution problems. In particular, the former East Germany has the highest sulphur dioxide emission rates per person in the world. The following methods could reduce the currently unsustainable levels of waste production in most European cities:

Checkpoint 2

What does the term *sustainable* mean?

→ establish **reduction targets** such as on the emissions of CO_2
→ adopt **waste management plans** to re-use, recycle and recover materials
→ improve **monitoring** of waste sites
→ establish a **comprehensive list** of contaminated sites
→ **coordinate waste management strategies** across international boundaries
→ establish indices of environmental management.

There are certainly more records of pollution in developed countries, but there are increasing levels of pollution in NICs and in developing countries. Countries like Bangladesh have little money to invest in pollution control. Industries in such countries favour the use of cheap, inefficient energy resources, such as lignite and low-grade coal. **Multinationals** are often responsible for pollution in these countries, such as that produced in the Bhopal disaster in India (1984), and the impact of *maquiladora* development in Mexico.

Examiner's secrets

Refer to any local evidence you have, such as your local council's policy on recycling, energy saving, Agenda 21, and so on.
Use local case studies. If these include strategies for dealing with pollution, so much the better. Check with your local council. You also need a case study that has an international angle, e.g. River Rhine, North Sea.

Exam practice   answer: page 88

'Pollution is the price of progress.' Discuss. (30 mins)

Water pollution

Water pollution is a 'serious **ecological disaster** comparable in importance to the destruction of the tropical rainforests and desertification'.

Increases in freshwater pollution

There is widespread pollution by sewage, nutrients, toxic metals, industry and agricultural chemicals; the most widespread is from domestic sewage. Poor waste-water treatment and inadequate sanitation have resulted in an exponential increase in waste pollution. Water pollution in many countries intensified in the twentieth century due to:

→ industrialisation
→ urbanisation
→ deforestation for urban growth and agriculture
→ the damming of rivers
→ destruction of wetlands
→ mining and industrial development
→ agricultural development
→ increased energy consumption.

There are six major problems facing the world's fresh water:

→ eutrophication
→ acidification
→ toxic contamination
→ decline of water levels
→ accelerated siltation
→ extermination of ecosystems and biota.

There is some evidence that developed countries have passed through a number of **stages of water pollution**. Developing countries (LEDCs) have experienced fewer stages but are expected to follow suit.

Checkpoint 1

What are the main sources of water pollution?

Checkpoint 2

What does *siltation* mean?

The jargon

Eutrophication means nutrient enrichment.

The main pollutants

Pathogens

The most common source of pathogens is **organic matter** from:

→ domestic sewage and municipal waste
→ industrial effluents, e.g. tanneries, paper mills and textile factories
→ storm runoff
→ land fills
→ agricultural areas.

This organic matter includes faecal material, viruses, bacteria and other organisms, as well as a wide variety of carbon compounds. Water-borne infections include schistosomiasis, hepatitis A and dysentery.

In the USA, 16 billion disposable nappies are dumped in landfill sites each year, and are a major source of concentrated pathogens.

Nutrients and eutrophication

Large concentrations of inorganic nutrients such as fertiliser can overload natural systems. In the UK, a major problem results when nitrates from agricultural areas percolate into the groundwater.

Many European rivers have extremely high levels of nitrogen and phosphorus – up to fifty times the natural background levels. This overloading causes **eutrophication**. This can create algal blooms leading to oxygen depletion and a decline in biological diversity. World fertiliser use is increasing rapidly, especially in developing countries. High levels of nutrient enrichment are also caused by seepage of water from septic tanks and pit latrines.

Heavy metals

Pollution by heavy metals occurs in a number of ways:

→ processing of ores and metals
→ industrial use of metal compounds
→ leaching from domestic and industrial waste dumps
→ mine tailings
→ contaminated bottom sediments
→ lead pipes.

The effect of heavy metals tends to be regional or local rather than global. However, they can affect areas downstream or downwind; for example, the Rhine Basin supports 40 million people and 20% of the world's chemical industry. Until the 1970s the Rhine was severely polluted but it has improved due to better waste-water treatment and the replacement of certain metals in industrial processes.

> **Check the net**
>
> Visit the World Health Organisation at:
> www.who.int

> **Grade booster**
>
> A contrast between a developing country and a developed country, with some reference to the model of development and pollution, will improve your chance of success.

> **Exam practice**   answer: page 88
>
> What are the causes and consequences of freshwater pollution? (30 mins)

Atmospheric pollution: acid rain

Check the net

Use the Friends of the Earth website www.foe.co.uk/ to find out current information on acid rain.

Acid rain – or **acid deposition** – is the increased acidity of rainfall and dry deposition, as a result of human activity. Rain is naturally acidic, owing to carbon dioxide in the atmosphere, with a pH of about 5.6. 'Acid rain' can be as low as 3.0.

Acid rain

Checkpoint 1

What is meant by (a) *acid rain* (b) *dry deposition*?

Rain has become more than usually acid because of air pollution. Snow and rain in north-east USA have been known to have pH values as low as 2.1. In eastern USA as a whole, the average annual acidity values of precipitation tend to be around pH4. As a general rule, sulphur oxides have the greatest effect, and are responsible for about two-thirds of the problem. Nitric oxides account for most of the rest. However, in some regions, such as Japan and the west coast of the USA, the nitric acid contribution may well be of relatively greater importance. Although emissions of SO_2 are declining, those of NO_x are increasing – partly as a result of increased car ownership.

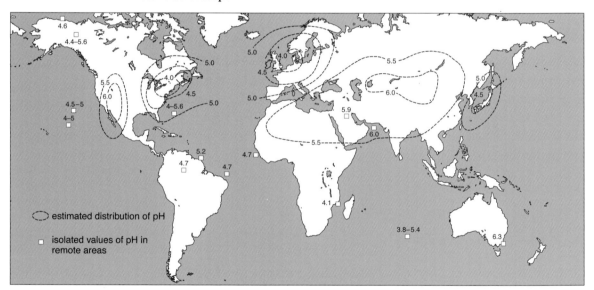

Checkpoint 2

What are the following chemicals: SO_2 and NO_x?

The **pH scale** is used as a measure of acidity or alkalinity: 7 is neutral, less than 7 is acidic and more than 7 is alkaline. The pH scale is **logarithmic**, so a decrease of one pH unit represents a tenfold increase in acidity. Thus pH4 is ten times more acidic than pH5.

The major causes of acid rain are the sulphur dioxide and nitrogen oxides produced when fossil fuels such as coal, oil and gas are burned. Sulphur dioxide and nitrogen oxides are released into the atmosphere where they can be absorbed by the moisture and become weak sulphuric and nitric acids, sometimes with a pH of around 3. Most natural gas contains little or no sulphur and causes less pollution.

Checkpoint 3

What is meant by a logarithmic scale?

Coal-fired power stations are the major producers of **sulphur dioxide**, although all processes that burn coal and oil contribute. Vehicles, especially cars, are responsible for most of the **nitrogen oxides** in the atmosphere.

Dry deposition typically occurs close to the source of emission and causes damage to buildings and structures. **Wet deposition**, by contrast, occurs when the acids are dissolved in precipitation, and may fall at great distances from the source. Wet deposition has been called a 'trans-frontier' pollution, as it crosses international boundaries.

Acidification has a number of effects:

→ buildings are weathered
→ metals, especially iron and aluminium, are mobilised by acidic water, and flushed into streams and lakes
→ aluminium damages fish gills
→ forest growth is severely affected
→ soil acidity increases
→ there are links (as yet unproven) with the rise of senile dementia.

The effects of acid deposition are greatest in those areas that have high levels of precipitation (causing more acidity to be transferred to the ground) and those that have base-poor (acidic) rocks that cannot neutralise the deposited acidity.

The solutions

Various methods are used to try to reduce the damaging effects of acid deposition. One of these is to add powdered limestone to lakes to increase their pH values. However, the only really effective and practical long-term treatment is to curb the emissions of the offending gases. This can be achieved in a variety of ways:

Checkpoint 4

Why do some industrialists deny that acidification is due to industry?

→ by reducing the amount of fossil fuel combustion
→ by using less sulphur-rich fossil fuels
→ by using alternative energy sources that do not produce nitrate or sulphate gases (e.g. hydro- or nuclear power)
→ by removing the pollutants before they reach the atmosphere.

However, while victims and environmentalists stress the risks of acidification, industrialists stress the uncertainties. For example:

→ rainfall is naturally acidic
→ no single industry/country is the sole emitter of SO_2/NO_x
→ different types of coal have variable sulphur content
→ cars are also a source of this type of pollution.

Exam practice   answers: page 88

1   Describe the pattern of the most acidified areas in the figure on page 86.
    (10 mins)
2   Explain the causes of acid rain (20 mins)

Examiner's secrets

Support your answers with facts, figures and case studies rather than general points.

Answers
Pollution

Principles of pollution

Checkpoints

1 Volcanic eruptions that can produce ash and gases.
2 *Sustainable* means an ecosystem or resource should be managed to continue naturally and not be destroyed.

Exam practice

You should use case studies to back up your argument. In general terms, pollution increases with population growth and with economic development. This is clearly shown in the model of air quality and levels of economic development. Less wealthy countries, such as Bangladesh, have poorer economies and investment in pollution control is minimal. Instead, such countries favour industrialisation and the use of cheap, inefficient energy resources, such as lignite and low-grade coal, as sources of energy. By contrast, rich countries, such as the UK, which may have gone through a process of deindustrialisation, have the capital and the technology to tackle air pollution. There is certainly more pollution in developed countries, but there are increasing levels of pollution in NICs and in developing countries. Many are related to the activities of multinational companies.

However, there are notable examples of countries where economic progress has occurred without a rise in pollution – Singapore is the country that comes to mind. The insistence of its government on anti-pollution measures, such as banning the use of chewing-gum, is a case in point. It has developed economically but not at the expense of the environment.

Water pollution

Checkpoints

1 The main sources of water pollution are from agriculture, industry and domestic sewage.
2 *Siltation* is the increase in silt going into water causing lack of clarity and, in the case of reservoirs, reducing the water storage capacity.

Exam practice

The causes of freshwater pollution include:
- urbanisation, industrialisation and intensification of agriculture
- deforestation for urban growth and agriculture
- the damming of rivers
- destruction of wetlands
- mining and industrial development
- agricultural development
- increased energy consumption.

The consequences of freshwater pollution include:
- eutrophication
- acidification
- toxic contamination
- decline of water levels
- accelerated siltation
- extermination of ecosystems and biota.

An essay should describe and explain causes and consequences, and use case studies to back up narrative.

Atmospheric pollution - Acid rain

Checkpoints

1 (a) *Acid rain* is rain that has become more acidic because of human action.
   (b) *Dry deposition* is particles of pollution that fall like dust.
2 SO_2 is sulphur dioxide; NO_x is nitrous oxides.
3 One that goes up by the power of 10 e.g. $10^3 = 1000$, $10^4 = 10\,000$, $10^5 = 100\,000$
4 If industry is accepted as the cause of acidification then there is the implication that the polluter should pay for damage or at least operate a cleaner system.

Exam practice

1 The most acidified areas are in the industrialised nations such as those in Western Europe and east Asia, and the USA. However, on a local scale, pollution is not in the main coal-producing and industrial areas. For example, within Europe, Scandinavia and Eastern Europe are very badly affected and it is thought that the sources of the contaminants are Britain and Germany respectively.
2 The major causes of acid rain are the sulphur dioxide and nitrogen oxides produced when fossil fuels such as coal, oil and gas are burned. These become weak sulphuric and nitric acids, sometimes with a pH of around 3. Coal-fired power stations are the major producers of sulphur dioxide, although all processes that burn coal and oil contribute. Vehicles, especially cars, are responsible for most of the nitrogen oxides in the atmosphere. Dry deposition typically occurs close to the source of emission and causes damage to buildings and structures. Wet deposition, by contrast, occurs when the acids are dissolved in precipitation, and may fall at great distances from the sources. Wet deposition has been called a 'trans-frontier' pollution, as it crosses international boundaries.

Revision checklist
Pollution

By the end of this chapter you should be able to:

1	Describe the main types of pollution.	Confident	Not confident **Revise** page 82
2	Outline ways of managing pollution.	Confident	Not confident **Revise** pages 82–83
3	Describe and explain the interrelationships between pollution and economic development.	Confident	Not confident **Revise** page 83
4	Describe and account for the increase in freshwater pollution.	Confident	Not confident **Revise** page 84
5	Describe the main pollutants.	Confident	Not confident **Revise** page 85
6	Define acid rain.	Confident	Not confident **Revise** page 86
7	Describe and account for causes, consequences and potential solutions to the problem of acid rain.	Confident	Not confident **Revise** pages 86–87

Population

Population is fundamental to Human Geography and underpins the studies of Settlement, Economic Change and Sustainability. The nature of the specifications does mean that population is studied at both A2 and AS. The emphasis is on understanding the nature of and components of population change. Population Geography focuses on the issues caused by both natural growth (over population) and decline (the impact of pandemics such as AIDS). In addition, the effects of migration on both the areas of origin and reception regions is an important issue this century, not least because of debates surrounding the impact of refugees and asylum seekers on affected countries.

A new major contribution to your geographical studies are the issues surrounding gender, whether attitudes to gender resulting in imbalance between the sexes, or the economic opportunities available to men and women. There are many demographic challenges facing countries in this century no matter where people settle to live and work. You will be expected to understand these challenges and why different people respond to them in different ways.

Exam themes

- Components of population structure and change

- The causes and consequences of migration

- The issues arising from migration of refugees and asylum seekers

- Imbalances in gender roles and their impact on the economy and society

- Population policies and the impact of aging societies

Topic checklist

	Edexcel		AQA		OCR		WJEC	
	AS	A2	AS	A2	AS	A2	AS	A2
Population: demography	O		O		●	O		
Migration	O		O		●	O		
Refugees and asylum seekers	O				●	O		
Gender issues		●					O	
Demographic challenges	O		O		●	O		

Population: demography

Check the net

The following websites can assist your studies of demography: www.populationconcern.org.uk, www.census.gov, www.prb.org and www.unfa.org. Those studying WJEC, and indeed all of you, may consult the Learning Grid for Wales at www.ngfl-cymru.org.uk/vtc/demographic_trans/eng/introduction. It is in English!

Population geography explains the location, number and changes in the people of an area over time. It is a dynamic study because people are born, die and move. We measure population in censuses, which are inevitably always out of date. The study of the components of population is called *demography*.

Distribution and density

→ 75% of the world's population lives on 13% of the land area.
→ All large concentrations are in the northern hemisphere between 10° and 55°N, with the exception of parts of South East Asia.
→ In 1980 the world population was 4,4301.1 million. By 2002 it was 6,198.5 million, even though the growth rate averaged 2.1% per year. Growth is not evenly spread.

The components of population change are summarised below:

INPUTS	PROCESSES	OUTPUTS
births	natural increase	deaths
+	+	+
immigrants	migration	emigrants
=	=	=
births and immigrants	total population change	deaths and emigrants

Action point

Can you draw sketch maps to show the distribution and density of population in a country that you have studied?

The jargon

Birth rate is the number of live births per 1000 people in a year and is a measure of fertility. *Death rate* is the number of deaths per 1000 people per year and is a measure of mortality.

Examiner's secrets

Make sure that you can define precisely the other types of demographic data such as fertility rate, life expectancy, infant mortality rate, dependency ratio.

Checkpoints

Look at the table. Where are the sparsely peopled regions and why is their population density so low? Can you relate the figures to the demographic challenges on pp9–9?

	Population Density per km^2	Crude death rate per 000	Crude birth rate per 000	Forecast % annual growth rate 2002-2015	Dependency ratio 0-15 as % working population	Dependency ratio 65+ as % working population
Singapore	6826	5	11	1.1	0.3	0.1
Bangladesh	1042	8	28	1.5	0.3	0.2
India	353	9	24	1.2	0.5	0.1
Sri Lanka	293	6	18	1.1	0.4	0.1
UK	246	10	11	0.0	0.3	0.2
Germany	236	10	9	–0.2	0.8	0.1
Italy	196	11	9	–0.3	0.2	0.3
Nigeria	146	17	39	1.9	0.8	0.0
Uganda	125	18	44	2.4	0.2	0.2
France	108	10	29	0.3	0.3	0.2
Ghana	89	13	44	1.7	0.8	0.1
Kenya	55	16	35	1.4	0.8	0.0
Burkina Faso	43	19	43	2.1	0.9	0.1
USA	37	9	14	0.8	0.3	0.2
Zimbabwe	34	21	29	0.6	0.8	0.1
South Africa	32	20	25	0.3	0.5	0.1
Brazil	21	7	19	1.1	0.4	0.1
Russian Fed.	9	15	10	–0.5	0.2	0.2
Canada	3	7	11	0.5	0.3	0.2
Australia	3	7	13	0.8	0.3	0.2

Demographic information

The figure below shows the demographic transition model.

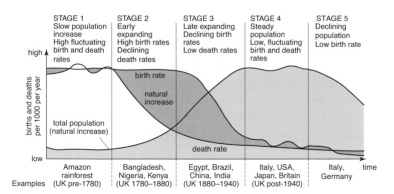

Population changes due to the interaction of births and deaths 2002

The population of the UK in 2006 was 60.6m. In 2016 the total is forecast to be 65 m of which 2.3 m (53% of growth) will be natural increase and 2.1 m (47% of growth) migration. Remember that all population forecasts are hard to predict due to fluctuations in migration.

Migration is the movement of people from one place to another and can be permanent or semi-permanent. It adds to or takes away from natural change. Circulation describes our daily commuting, movement for holidays, and nomadism.

Grade booster

You will impress the examiner if your example of a country at a stage of the demographic transition is not from a textbook.

The jargon

Natural change is the difference between birth and death rates. Overpopulation is reached when an area cannot support its population. Incomes are low or declining, living standards are declining, there is high unemployment and people are migrating away.

Examiner's secrets

Always have a bank of quickly drawn population pyramids to illustrate answers about population structure.

Examiner's secrets

You do not have to be totally precise when you quote population data. Examiners want statistics but are happy if you are quoting close to the right figures. Do not use phrases such as 'as in Italy' without supporting evidence.

Exam practice   answers: page 102

1   Why is the world's population growth varying between continents and countries? (15 mins)

2   Describe the characteristics of the population pyramids shown below. Suggest a type of settlement or country which each one might represent. (10 mins)

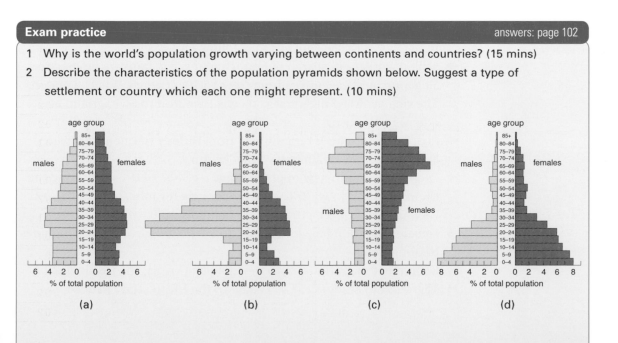

Migration

Migration is the cumulative result of many individual decisions. It can be international, or interregional within a country, or between places within a region.

Action point

Have you got a classification of the types of migration either in diagrammatic or note form?

Check the net

See www.refugeecouncil.org.uk, www.worldrefugee.com, www.ilo.org and www.statistics.gov.uk. Note that many websites such as www.migrationwatch.org do present a particular viewpoint and may be biased. You should always ask yourself why they are putting forward a particular opinion and why you agree or disagree with that viewpoint.

Examiner's secrets

The examiner will look for the names of people associated with theories. For instance, Lee's model is about intervening opportunities.

Why people migrate

→ Push–pull factors (which can be both pull and push factors).
→ Perceived quality of life – the pull of London or even the catchment of a good school.
→ Harsh environments – Sahel, or crofting in the Hebrides.
→ Population pressure – Bangladesh or the crowded nature of cities pushing people to the countryside.
→ Marriage, including arranged marriage.
→ Forced (see refugees).
→ Labour migration or economic migrants – Polish plumbers.
→ Persecution and political factors (see refugees).
→ Migrants act in economic self-interest, weighing up costs and benefits (Todaro 1971).
→ The inevitable outcome of capitalism (Marxist view).

Why and how do people migrate according to Ravenstein?

→ The greater the distance, the fewer people migrate = Zipf's inverse law of migration 1949.
→ The bigger the pull, whether it be in numbers or wealth, the greater the flow of people.
→ Some people migrate in stages – intervening opportunities delay or even stop the full move = Stouffer's Theory of Intervening Opportunities.

The international migrant

→ Migrants are often male and young, e.g. those who migrated from the Caribbean to the UK in the 1950s, or Poles and other East Europeans to Germany and the UK since EU expansion.
→ The less skilled migrate in search of opportunities, e.g. the Irish migration to the USA in the nineteenth century or Pakistanis to Dubai in the 1990s. Families may follow on.
→ Skilled workers in search of opportunities, e.g. bankers moving to Singapore and executives from New York to Canary Wharf, London. The rich avoiding high taxes – Lewis Hamilton to Switzerland and others to Monaco.
→ People migrate from tyranny, e.g. from Hitler's Germany, or from Kosovo in 1999, or Zimbabwe this century.
→ People migrate to new opportunities, e.g. 40% of those entering the UK in 2006 had jobs (27%) or were looking for jobs. In 2006 29% of UK immigrants were attracted by London (2000 = 43%).
→ Female migration has grown, e.g. domestic help from the Philippines to Hong Kong.
→ Retirement migration (e.g to the Spanish Costas, the Dordogne).
→ Migration to study – 157,000 came to the UK to study in 2006.

Migration at a national level

→ 19th century–people moved to the UK coalfields from rural poverty in search of work: the process of urbanisation. This is still a feature of cities in the LEDCs.

→ 20th century movement towards London and the south-east.

→ Since the 1930s people have moved to the outer edges of cities: the process of suburbanisation.

→ Since the 1960s there has been a net movement of people away from urban areas back to small towns and villages: the process of counterurbanisation.

→ Retirement migration to the south coast and to south-west England.

→ There is some migration back into cities – reurbanisation – mainly by young, high-flying, single people and the very rich (e.g. Islington, Cardiff Docklands).

Action point

Do you have case studies of international migration and its causes and effects, e.g. nineteenth-century Irish to the USA, South Asians to the Middle East, Estonians and Poles to the UK?

Checkpoint 1

Where are people coming from when they legally enter the UK and where are UK citizens migrating to? How have these flows changed in the past 50 years?

Checkpoint 2

What are the economic effects of international migration on a receiving area in the UK?

What are the effects of migration?

The exporting area: LEDCs MEDCs	The receiving area:
loses its breadwinners	pressure on services and utilities such as water supply
loses skills such as nurses	more and possibly cheaper labour available because willing to work for less. Impact of gangmasters
becomes more elderly and loses next generation. Dip in birth rate	a rise in birth rate
loses its gender balance, with population pyramid skewed due to loss of young adults and males	a younger and possibly gender-imbalanced population structure
loses services if migrants are the best qualified – 20% of new jobs in UK filled by immigrants are in public services and 20% in banking	facilities and services geared to the dominant group, e.g. mosques, temples, specialist retailing and travel agencies
suffers from land being abandoned	increased demand for housing and education
	overcrowding and environmental effects of more people
may see a loss of elderly population due to retirement migration	an influx of immigrants can cause other groups to leave the area
gains from remittances	education provision may have to cater for young people whose first language is not that of the host society
	the spread of diseases such as AIDS is partly brought about by migration. This increases the number of orphans and decreases the working population, especially in LEDCs
	pressures on countries on the migration route e.g. West Africans in Libya waiting to attempt to move to EU
	jobs filled that host citizens do not wish to fill

Migration is not limited as an issue to the most developed countries.
In Japan there are 250,000 'Nikkeijin' from Brazil and Peru.

The jargon

Remittances are the money sent home by migrants living abroad to their family in the region/country that they left. It can be in kind, such as cars taken home to help set up a taxi firm.

Exam practice   answer: page 102

Attempt to justify a classification of migration. (10 mins)

Refugees and asylum seekers

In 1951 the UN Refugee Convention established the rights of refugees.

Why are people refugees?

→ Displaced by conflict – Iraq (12,000 in Syria), Afghanistan.
→ Displaced by prejudice e.g. Jews in 1930s to London.
→ Displaced by internal conflict – Darfur (29400 in Chad 2008), Chechnya, Aceh, Indonesia, drug gangs in Columbia.
→ Displaced by environmental disasters – Aceh tsunami.
→ Civil war – Democratic Republic of Congo (exacerbated by access to minerals such as diamonds) where 2.5 million people are affected.

Checkpoint

Why are there special border patrols shown on the map?

Key migrant routes from Africa to Europe

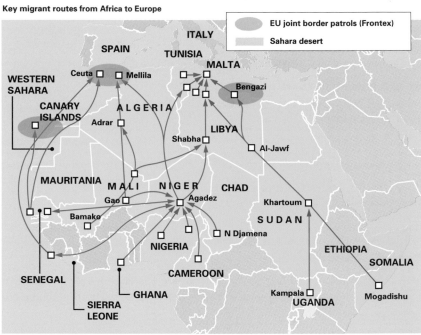

Source: Frontex, Europol and ICMPD

Refugees, asylum-seekers, internally displaced persons (IDPs), returnees (refugees and IDPs), stateless persons, and others of concern to UNHCR by country/territory of asylum, end-2007

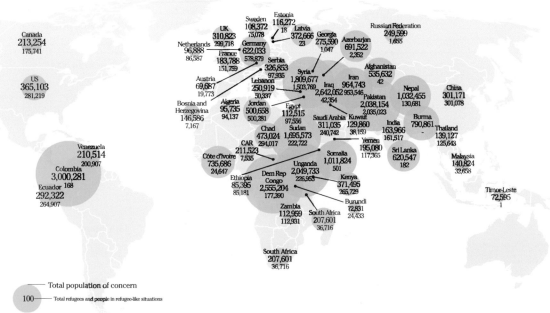

Darfur

The conflict has killed up to 300,000 and has driven 2.7 million from their homes, some of them to Chad. In 2003, a conflict between the rebels in the western region and the Sudanese government was fuelled by competition between Arab nomads and African farmers for scarce water and land, after years of drought. Herders whose animals were allowed on farm land because of the fertilising manure they dropped were stopped by the farmers. Some herders were recruited by the Janjaweed (Arab militias) who brutally oppressed the farmers who fled as refugees.

Asylum Seekers

Globally there were approximately 773,000 asylum seekers in 2005. The UK receives 1 in 10 of all applications for asylum.

Rank	Worldwide Countries of origin 2005	Number	Rank	UK Countries of origin of applicants 2001	Number
1	Democratic Republic of Congo	55963	1	Afghanistan	9000
2	El Salvador	45205	2	Iraq	6800
3	Myanmar	41131	3	Somalia	6500
4	Iraq	34441	4	Sri Lanka	5500
5	Guatemala	31850	5	Turkey	3600
6	Somalia	30467	6	Iran	3400
7	Columbia	19754	7	Former Yugoslavia	3200
8	China	19403	8	Pakistan	2700
9	Serbia, Montenegro	18132	9	China	2300
10	Zimbabwe	17326	10	Zimbabwe	2100
11	Pakistan	16458		**UK 2001**	71,700
12	Rwanda	15880		**Total 2007** (UNHCR data)	40,200

Between 76% and 58% of all asylum applications (dependent on the year in question) are rejected, so the overall impact is less than the sensationalists claim. However, some do avoid repatriation. In 2004 applications to the UK from asylum seekers and their dependents were only 0.7 per 1,000 of the population compared to 2.6/000 in Sweden, 3.1/000 in Malta, 1.9/000 in Switzerland, 1.7/000 in Belgium and 1.2/000 in Ireland.

The jargon

Asylum Seeker A person attempting to gain entry to a country by claiming refugee status. Their life is endangered if they remain in their home country. Persecution, famine and hardship are frequent reasons.

The jargon

Economic migrant Person seeking to migrate in order to obtain better economic circumstances through work in the destination country. Many asylum seekers and refugees are desperate economic migrants seeking to find a way into countries with migration controls. The EU permits free movement of labour within its boundaries, hence the economic migrations from East to West Europe.

Grade booster

When writing about contentious issues make sure that your response is free from bias. Some web sites can be very biased.

Exam practice   answer: page 102

Study the table and map. What are the similarities and differences between the global origins of asylum seekers and asylum seekers to the UK?

Gender issues

Women's roles in society have demographic, economic and social causes and effects, including attitudes to work and child rearing, attitudes to female children, and attitudes to migration. There can be imbalances in gender roles, such as in the north-east of England where there has traditionally been less work for women because the main employment was coal mining and shipbuilding. In contrast, Lancashire textiles had a tradition of female work.

Migration and gender

Migration affects sex ratios in the areas of origin and destination.

→ The ratio of males to females entering the USA was 2.5 to 1 in 2001.
→ Areas near the frontier with Mexico where illegal migration is common are characterised by few women.
→ The boatloads of Africans attempting to reach Europe from Libya and making the hazardous crossing to Lampedusa (Italy) and Malta are overwhelmingly male.

Other factors that may affect gender balance in an area

→ War deaths reduce the number of young males.
→ Incarceration of a disproportionately large group of males from a population can also affect the M/F ratio – in some US cities this is noticeable in black communities.
→ Longevity. Females tend to live longer on average.

Life expectancy at birth UK	Males		Females	
	1960	2005	1960	2005
	67.1	76.9	73.7	87.1
Average number of years until death at age 65 in 2002	16 years		19 years	

Demographic impact of educating girls in LEDCs

In 1999, 130m schoolchildren could not read or write, according to UNICEF, and 73m of those were girls. In sub-Saharan Africa, boys outnumber girls in secondary schools by 5 to 1. The figure below shows the demographic impact of educating girls.

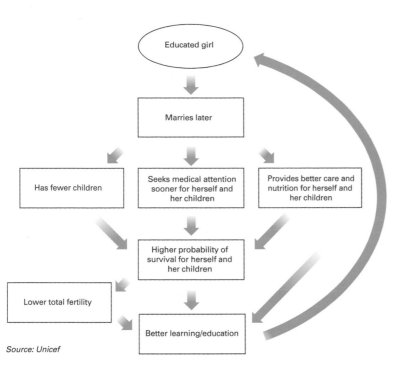

Source: Unicef

Gender and labour force

	Female % of labour force 1999	% females in agriculture	% in industry	% in service sector
Saudi Arabia	15.5	25	5	70
Algeria	27	69	6	25
Japan	41.3	6	23	71
Ghana	50	57	14	29
UK	43.9	1	13	86

The model above shows how feminisation of the workforce occurs most in the lower paid and casual labour markets, where there is an increasing insecurity of livelihood. In Africa and parts of South Asia female labour has been concentrated in agriculture, whereas in the rest of Asia women are found almost equally in all sectors and increasingly in the service sector. The demographic transition has given more women the opportunity to work and to be better educated, enabling them to be employed in higher status jobs. However, globalisation has resulted in some 'female' jobs becoming more informal and casual as home working is used.

Gender concentrations

Gender concentrations can occur in cities and were first identified by Park and Burgess in the 1920s: they called them 'moral regions'. In San Francisco 18th and Mission St is the home of female prostitutes, Polk Street is known for male prostitutes, the Castro is the district where gay men tend to live, and Valencia Street is at the heart of the lesbian community.

Society's attitude to gender is also an important field of demographic study.

Exam practice   answer: page 103

What are the demographic and economic implications of the data in the table? (15 mins)

Action point

Make some notes on the different attitudes to women in MEDCs and LEDCs and how this affects changes in population numbers, and roles in the economy and in society. The table will help you.

The jargon

Fertility transition is the move to lower birthrates in LEDCs as the result of female education, which impacts on child mortality and contraception.

Check the net

www.worldbank.org is an excellent source of demographic data including data on life expectancy.

Grade booster

If you are studying sociology, psycology or media studies, do not hesitate to use examples that you have studied for those disciplines. For instance, in 2008 there were comments in the media on the 'glass ceiling' which you could use in some answers.

Demographic challenges

Check the net

www.populationconcern.org.uk
www.unAIDS.org
www.ageconcern.org.uk
www.helptheaged.org.uk

The jargon

Overpopulation occurs when the ability of an area to support its population by feeding the population or by providing jobs to enable people to live is exceeded.

The jargon

When there is exponential growth of a population this is called a population explosion.

Demographic challenges relate to many aspects of human geography and especially to development and health and welfare. The most important issues are population control, the impact of population decline, the impact of migration, ageing and life expectancy.

Overpopulation and population control

In 1900 the world's population was 1.5 billion people and by 2000 it was 6 billion. LEDCs are especially prone to higher population growth rates although disease and consequent higher death rates do limit the rate of growth a little. This is due to **life expectancy** rising, birth rate not falling and health control improving. A few countries are also trying to increase population.

Types of policy

Expansionist policies

→ Restrict availability of birth control, e.g. Ireland and Ghana.
→ Make abortion illegal, e.g. Guernsey.
→ Give larger family allowances to those with more children, e.g. France.
→ Support for working mothers, e.g. France, UK.
→ Improve health care.
→ Restrict roles for women in society, e.g. Saudi Arabia.

Control policies

→ Promote birth control, e.g. Singapore.
→ Permit abortion.
→ Encourage sterilisation, e.g. India.
→ Control the number of children in a family, e.g. China's one-child policy (1980s).
→ Governmental advertising campaigns, e.g. India.
→ Health education in schools e.g. Netherlands.

Project tip

If you live in or near an area with a higher than average elderly population (e.g. Bognor, Budleigh Salterton, Worthing) see if you can identify the impact that this has on the provision and location of local services.

Checkpoint

In what way are the elderly a different issue in LEDCs?

Examiner's secrets

Be careful about being ageist. Many examiners are in late middle age and approaching retirement and you don't want to offend them!

Ageing in the UK

In 2006 16% of the UK population or 9.7 million people were over 65. Those over 85 are 5.9% of the population. By 2031 the over-65s are expected to rise to 22% and in 2050 (when many of you will be retired) to 29%. There were 4.7m persons over 75 in 2006 and this will rise to 8.2m in 2030. The elderly are dominated by females with a ratio of 234 females to 100 men among those over 80. In 2007 pensioners exceeded the number of children.

Distribution of rising life expectancy

→ Retirement migration to the south-west and south coast of England.
→ Out-migration of younger age groups from central Wales.
→ Older people are left behind in inner cities because of low income and by out-migration of the young.
→ In all areas, numbers swelled by the 1940s baby boom are now reaching retirement.

Effects on policies

→ Restriction of conversion of properties to care and nursing homes, e.g. Bournemouth.
→ Diversification of local economies to counteract immigration of old.
→ Companies set up to provide homes for the elderly e.g. McCarthy & Stone.
→ Where the elderly vote is large, this is likely to be reflected in policies, e.g. law and order.
→ Health provision pressures, ranging from care to geriatric beds.
→ Policies to employ the fit elderly, e.g. B&Q.
→ Raising retirement age because of pension costs.
→ Increased social services for the elderly, and access to retailing geared to the elderly.

Impact of declining birth rates

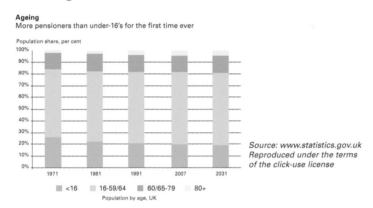

Ageing
More pensioners than under-16's for the first time ever

Population share, per cent

Source: www.statistics.gov.uk Reproduced under the terms of the click-use license

Population by age, UK

■ <16 ■ 16-59/64 ■ 60/65-79 ■ 80+

Grade booster

When talking about places, try to add the detail that suggests you have studied the area in detail.

→ Declining UK births 1990–2002 has resulted in lower demand for primary school places 1995–2007 and secondary school places until 2012. This has led to school closures and mergers.
→ Fewer people entering the workforce since 2003, whereas more retiring.
→ Immigrants replacing workforce shortages.
→ Retirees being persuaded to stay longer at work.
→ In some countries declining births have resulted in population decline since 1990 (e.g. Hungary). In others, emigration is predicted to have a similar effect (e.g. new EU countries).

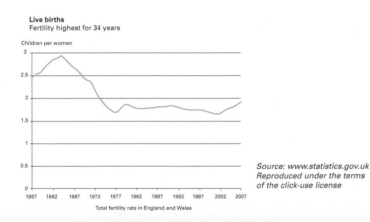

Live births
Fertility highest for 34 years

Children per women

Source: www.statistics.gov.uk Reproduced under the terms of the click-use license

Total fertility rate in England and Wales

Exam practice   answers: page 103

1 Discuss the impact of the changing structure of the population on the provision of services in a named country. (15 mins)
2 What are the processes involved when migration leads to the formation of ghettoes? (25 mins)

Examiner's secrets

If a question says a 'named country' make sure that you name the country. Note that Africa is not a country, and using such an example will not gain you marks.

Answers
Population

Population: demography

Checkpoint

Highest densities in very small states and lower in the very large states in the less hospitable parts of the world. Low also where high living standards partly due to high resource availability. You need to be very specific for each country. Challenges: overpopulation = where growth rate high and dependency of young high e.g. Burkina Faso and Uganda and possibly Nigeria and Ghana; ageing = low death rate and higher old age dependence together with very low growth rates e.g. Italy, Germany, Russian Federation and UK.

Exam practice

1 You could explain variations between countries by the stage of the demographic transition of that country. This would lead you to illustrate variations in birth and death rates, and life expectancy with approximately correct figures that you have memorised. Variations can also be explained by the availability of resources. Population policies might also come to bear on some countries, as will religious beliefs, and especially Islam on birth rates in some Arab countries. Within a country, demographic changes are less important on the whole, but migration does become an important factor. Places that have attracted a younger population will grow more rapidly than those places that attract older people.

2 (a) A larger number in early adulthood. This could be a place with married workers moving in or subject to immigration from overseas (see Migration on pp 94-95). It might be an established settlement in a country with a low birth rate.

(b) Is a community with a large influx of males – many resource frontier mining settlements are like this. It is also characteristic of a place with a high migrant labour force. If it were a country, the adult part of the pyramid is characteristic of UAE, although Arab countries tend to have a high birth rate.

(c) A retirement community very much in the USA model – many communities in Arizona such as Scottsdale (Phoenix) have this age structure. No UK retirement areas have reached this state although wards in some south coast towns have a similar structure.

(d) This graph is characteristic of a town with a large number of both sexes in childhood and early adulthood. Therefore it is a young population very characteristic of a new town style of settlement.

Migration

Checkpoints

1 Migration is from the Caribbean (1950s), from S Asia (1970 onwards), the New Commonwealth (115,000 in 2006) and from the EU, especially the new member states since 2000. (London has a population of 300,000 French citizens, enough to make it the 4th largest French city in the world!) 400,000 emigrated in 2006, of whom 207,000 were British and 194,000 were foreigners returning home after at least a year in UK. Emigration of British citizens was to Australia, New Zealand (Commonwealth) – continuing flow over last century, USA (employment partly result of globalisation), France and Spain (retirement migration and some working) – a more recent flow although destinations changed - Costa Brava (1960s) Costa del Sol (post 1980).

2 Concentration into some areas – ghettos, schooling with English as a second language - a Primary school in Sutton has 30% of students with English as a second language and there are 15 different first languages. Pressures on local authorities to provide shelter, support issues, issues to do with development of cultural facilities such as temples and mosques (catholic churches in many cities have one Sunday service in Polish). Retail specialisms such as Polish delicatessen and travel agencies specializing in travel back to countries of origin. Airlines set up routes to assist newcomers e.g. Bournemouth to Kracow to cater for 20,000 Poles in Bournemouth. Newspapers in the home language e.g. Polski Echo in Bournemouth.

Governments help by providing human and physical resources, by legally combating any racism that might occur, by attempting to disperse population, by producing documents in a variety of languages. Local migrant support networks and social facilities for the newcomers normally arrive.

Exam practice

Your answer should include a classification; there are many in your textbooks. Don't forget to make passing reference to short term movement such as commuting. The second part of your answer must be some reasoning, with supporting examples, for the chosen classification. Justification implies that the answer should say why you prefer a more sophisticated classification rather than a simpler one.

Refugees and asylum seekers

Checkpoint

The flows on the map are mainly economic migrants either attempting to enter the EU illegally, or to claim refugee or asylum seeker status. The border patrols operated by the EU are because Ceuta and Melilla are enclaves of Spain in Morocco and are therefore part of the EU – under 4,000 people attempt this route now that the border is fenced and patrolled. Similarly, the Canary Islands are Spanish and have attracted people traffickers using ill-equipped boats (30,000 used this route in 2006). Many (circa 6,000) also crossed to mainland Spain. Bengazi was the point of departure to Malta, Lampedusa and Sicily although the point of embarkation moved westwards to avoid the patrols. Frontex (an EU/local police agency) now patrols between Libya and Italy.

Exam practice

Points you could make are: some countries in both lists and obviously for the same reasons such as wars, internal conflicts and civil war; the UK list includes countries that are nearby – distance decay plays a part; the UK Commonwealth links are strong; UK does not have significant applications from Latin and Central America; African countries generally absent except for Zimbabwe where in addition to the Commonwealth link there are the persecuted whites and opposition supporters fleeing Mugabe's regime; plus flight from hardship in Somalia (partly caused by civil strife).

Gender issues

Exam practice

We see rising life expectancy for both sexes but increasingly more for women. The implications are: more single-person households as partners die; because homes are occupied longer more houses are needed to cater for rest of growing population; more care homes as people live longer; a political force is emerging (the grey vote); more services to cater for the elderly (such as SAGA Insurance and Travel); potential use of the elderly as carers for children where both parents work; rising demand for health services and public transport. Most of these implications apply to both sexes although if women live longer, the severity of the implications will increase.

Demographic challenges

Checkpoint

The elderly are different because there are fewer of them, they are part of the extended family and they have a role in child care. Because of the AIDS pandemic many become the primary carers for their orphaned grandchildren. Care for the elderly will be within the family. They are invariably working even if it is only within subsistence agriculture.

Exam practice

1 A good way to approach this question would be to start with a series of population pyramids for a country (UK?) from 1901 to 2001. From these you would be able to show that there have been changed demands for services. The post-1945 baby boom led to demands for schooling and later for house-building and e.g. New Towns. More recently the slowing down of population growth and smaller family size have led to an ageing society. The issues of caring for and supporting the elderly, ranging from migration to the provision of suitable facilities, should be discussed with the support of examples which could come from almost any area of the UK.

2 The processes are those of clustering of peoples in areas where they feel at home with 'like' people. Immigrant groups tend to start in key low-rental areas and then spread outwards, often displacing the former population. Displacement is the result of both 'white flight' and the new people being the only ones interested in the area. Some have moved to the outer suburbs as they have become more established. Low-rent areas are also close to industrial areas and the airport.

Revision checklist
Population

1	Use approximate demographic information to illustrate the characteristics of countries at stages of the demographic transition.	Confident	Not confident **Revise** pages 92–93
2	Draw the five-stage Demographic Transition Model.	Confident	Not confident **Revise** page 93
3	Explain why people are migrating to specific regions such as the South East of the UK.	Confident	Not confident **Revise** pages 94–95
4	Explain why people migrate internationally.	Confident	Not confident **Revise** pages 94–95
5	Cite a detailed case study of the refugee problem.	Confident	Not confident **Revise** pages 96–97
6	Distinguish refugees and asylum seekers.	Confident	Not confident **Revise** pages 96–97
7	Explain what causes people to be refugees and asylum seekers.	Confident	Not confident **Revise** pages 96–97
8	List the factors that affect gender balance in different parts of the world.	Confident	Not confident **Revise** pages 98–99
9	Explain why the number of elderly people in Western Europe is growing so rapidly.	Confident	Not confident **Revise** pages 100–101
10	See both sides of many demographic issues and not display bias.	Confident	Not confident **Revise** pages 100–101

Settlement

The key themes of this section concern recent changes in the form and function of settlements and the economic, political and social issues that have come about as a result of these changes over time. In 2007, more of the world's population lived in cities than in rural areas. In the developed world rural settlements are rarely just agricultural. Urban areas have changed from being industrial centres and market towns into centres for the modern service economy. Towns and cities have acquired a distinct patterning of activities and social areas, and the processes which bring about these changes are a key focus of settlement geography. A key characteristic of settlements is that they are constantly changing, and possibly growing in extent, in ways that people may or may not appreciate. Therefore, understanding the reactions of decision makers, planners and the general public to changing settlements and why people and groups hold particular attitudes, values and opinions are key elements of your studies.

Exam themes

- Cities are constantly changing economically, socially and environmentally.

- Change is managed by governments and organisations, and people respond to the management of change.

- Rural settlements in MEDCs have changed to be more influenced by urban areas.

- Current issues in settlements are deprivation and crime.

- Much of peoples' image of changing settlements is based on their perception of environments.

Topic checklist

	EDEXCEL		AQA		OCR		WJEC	
	AS	A2	AS	A2	AS	A2	AS	A2
Urbanisation	O	●	O		O		O	
Regeneration and reurbanisation		●	O	●	O	●	O	●
The central business district	O	●		●	O		O	●
Social structures and diversity	O	●	O	●	O	●	O	●
Inner city issues	O	●	O			●	O	●
Retailing and services		●	O	●	O		O	●
The urban fringe	O		O			●	O	●
Rural settlement changes	O				O		O	
Leisure and recreation			O	●			O	●
Deprivation				●			O	●
Geography of crime		●	O	●	O		O	●
Perceiving environments								●

Urbanisation

Urbanisation is the term we give to the process of becoming urban. It is part of a model illustrating the processes of population movement in and around a city. Urbanisation differs in its nature between the continents and you should be able to distinguish urbanisation in MEDCs from that in LEDCs and LDCs.

The cycle of urbanisation

1. People leave the countryside for the cities, and population also grows naturally in the city.
2. People get better jobs and move or filter to better areas in the city – suburbanisation and urban sprawl.
3. People leave the city for the towns and villages beyond the city – counterurbanisation.
4. People yearn for city life and return to live close to the heart of the city – reurbanisation and gentrification.

In Europe the process was associated with the Industrial Revolution. It involved rapid expansion of urban areas, with people working in secondary jobs. Urbanisation was (and is) a social process because it changed a rural society into a more stratified one, where people were wage-earners who lived separately from the owners of the new factories. In the 20th century, urbanised areas became associated with employment in major corporations in the service sector and public sector employment.

Urbanisation in LEDCs and LDCs is happening in the 21st century and involves both a pull by the city and a push from the countryside.

→ Rapid demographic change is caused by migration and natural increase.
→ Socio-economic change is due to a shift from subsistence agriculture to a market economy.
→ Behavioural change due to the changing way of life.
→ Under-employment in the countryside is replaced by unemployment in the city – urbanisation without industrialisation. However, in the emerging Asian economies urbanisation is associated with industrial growth (China) and service industry growth (India).
→ Urbanisation led to primate cities – a city that has a disproportionate percentage of the national population.
→ It also leads to shanty towns/favelas/bustees because in-migration exceeds the rate of house building.

The jargon

Suburbanisation involves the *decentralisation* of people and employment from the centre of the city to the margins – it is often seen as the growth of a city.

Counterurbanisation was first noted in the USA by Berry. It is associated with the growth of tertiary and quaternary employment and involves people moving away from the city.

LEDC is a Less Economically Developed country designated by the UN and the World Bank.

LDC is a Least Developed Country designated as above.

Grade booster

Have worked-up case studies of urbanisation in different parts of the world ready.

Percentage of population living in urban areas, by region, 1960–2030

	1960	1980	2000	2020	2030
World total	32.8	39.1	46.7	55.1	59.9
Developed regions	58.4	68.7	73.2	77.8	80.8
North America	69.9	73.9	79.1	84.6	86.7
Europe incl. CIS	56.5	67.9	71.7	75.1	78.3
Oceania	66.6	71.2	70.5	72.3	73.8
Less developed regions	21.7	29.5	40.3	50.7	56.1
Least developed countries	9.5	17.4	24.7	34.4	40.9
Africa	18.7	27.6	36.20	45.3	50.7
Asia	19.9	26.3	37.1	48.1	54.1
Latin America & Caribbean	49.2	65.1	75.4	81.9	84.3

Source: *United Nations, World Population Prospects: The 2006 Revision and World Urbanisation Prospects* (New York, 2007)

The hierarchy of settlements

Megacities, often called **global hubs**, are the key centres in the modern world economy. London, New York and Tokyo are the major centres of banking, finance and investment. To these one might add the rising financial centres; Shanghai, Singapore (not mega in population) and Bombay. If you just take population of the urban area, the megacities are those listed on the right.

Conurbations are urban areas that have merged yet maintain different centres. Some were the product of the industrial revolution (e.g. the Ruhr) but they are also the product of the sheer size of adjoining cities – a mega-conurbation (e.g. 'Boswash', the US seaboard from Boston to Washington).

City is a large settlement dependent on commerce, manufacturing and service industries. The **city region** is the area served by a city which can be the journey to work region and journey to study area. The flexibility of modern transport enables city regions to overlap (e.g Fareham is in the city regions of Portsmouth and Southampton).

Towns are smaller urban areas without the range of facilities of cities.

Village is a settlement in rural surroundings which was once an agricultural settlement. Villages normally contain some services such as a church (or a mosque or temple), an inn and maybe a shop. 'Village' is frequently misused by estate agents to try and provide identity (and a higher price) to settlements engulfed by suburbia.

Hamlet is a small cluster of dwellings/farms in a rural area which lacks any services.

Exam practice
answers: page 130

1 What factors explain the distribution of the major cities shown in the table above? (15 mins)

2 List some of the similarities and contrasts between urbanisation in MEDCs and in LEDCs. (15 mins)

> **Checkpoint**
>
> Can you define a primate city? Name one.

> **Action point**
>
> Know your megacities
> World's largest urban areas, 2006 (millions)
> 1. Tokyo, Japan – 34.1
> 2. Mexico City, Mexico – 22.6
> 3. Seoul, South Korea – 22.2
> 4. New York, United States – 21.8
> 5. Sao Paulo, Brazil – 20.2
> 6. Mumbai, India – 19.7
> 7. Dehli, India – 19.5
> 8. Los Angeles, United States – 17.9
> 9. Shanghai, China – 17.9
> 10. Jakarta, Indonesia – 17.1
> 11. Osaka, Japan – 16.8
> 12. Kolkata, India – 15.5
> 13. Cairo, Egypt – 15.4
> 14. Manila, Philippines – 14.8
> 15. Karachi, Pakistan – 14.1
> 16. Moscow, Russia – 13.7
> 17. Buenos Aires, Argentina – 13.4
> 18. Dhaka, Bangladesh – 13.1
> 19. Rio de Janeiro, Brazil – 12.0
> 20. Beijing, China 11.9, and London 11.9

> **Action point**
>
> What makes the following places distinct; Portsmouth, Brighton, Norwich, Bradford, Cardiff, Tewkesbury?

> **The jargon**
>
> *Settlement* or *urban hierarchy* describes the classification of places using variables such as population size or status, or perceived importance. Most settlements in MEDCs are part of a continuum, an imperceptibly changing set of functions that overlap.

Regeneration and reurbanisation

Check the net

www.bristolregeneration.org.uk. You will find many more by Googling a city name and 'regeneration'.

The jargon

A greenfield site is a development on rural land. A brownfield site is where a new activity takes place on a site that has already been used and abandoned.

Links

This topic is also relevant to your studies of Retailing and Deprivation.

The jargon

Urban regeneration is the term created by 1999 Lord Rogers' Urban Task Force, to describe a wider process of environmental and social improvements.

Urban redevelopment means starting again on brownfield sites.

Urban renewal means working within the existing built form.

Action point

Do you know of examples in your area of brownfield sites? What was on them and how they are used now? Why did the change take place? Who gained and who lost out from the changes?

Reurbanisation concerns bringing people and jobs back to cities. The jobs are in new expanding service sector activities, such as the Canary Wharf financial district which employs young, affluent professionals, mainly in banking and finance, who often live nearby in the regenerated London Docklands. **Regeneration** is the planning policy term for the process of reviving employment and the social fabric of areas. Much regeneration is on **brownfield sites** in areas that have de-industrialised such as the 2012 Olympic Park in the Lea Valley, Cardiff Docklands, East Manchester and East Glasgow. It is associated with **urban renewal**.

Who is reurbanising?

→ Young and affluent people with no children.

→ Those for whom an address in an 'up-and-coming area' is a statement about their status.

→ Affluent working population needing to be close to work (Wapping, Chelsea and Kensington in London and Cardiff docklands).

→ Migrants from suburbs and beyond wanting to be close to services as they grow older.

→ Single-person households as product of divorce and relationship breakdown.

→ High-paid migrant workers on contracts – firms house these persons in desirable areas.

Why is regeneration necessary?

→ Deindustrialisation and closure of factories e.g. the Lower Swansea Valley.

→ Dereliction of sites and subsequent wasteland which can be toxic in some circumstances.

→ Can give the wrong impression to visiting business decision makers and tourists if dereliction is on a key route into the town or city.

→ Military establishments surplus to requirements.

Types of regeneration

→ Expanding the Central Business District into the Zone in Transition, e.g. London Bridge.

→ Rebuilding within the CBD to create more space in taller buildings – intensifying density of use, such as the London 'Gherkin' and the 'Shard'.

→ Reworking of former docklands for commercial and/or residential use, e.g. Canary Wharf, Gunwharf Quays, Portsmouth, Salford Quays.

→ Reworking former industrial sites for residential use, or for hotels (Malmaison hotels are frequently found in once abandoned buildings e.g. Manchester, and Leith).

→ Refurbishing and modernising areas of run-down housing often called gentrification e.g. Islington in 1990s.

→ Creating a museum to the lost activity e.g. Blaenavon's iron works (1788–1900 and coalmine closed in 1980).

→ Other cultural activities either in reused buildings such as Art Galleries (Baltic Centre for Contemporary Art, Gateshead), or purpose built buildings (Lowry Centre, Salford).

→ Moving in a key institution or organisation to attract others e.g. the Scottish Executive in Leith has attracted many companies, architects and legal businesses to the former dock area of Edinburgh. The Welsh Assembly is having the same effect in Cardiff Docks.

→ Major sports venues such as Eastlands, the stadium for the Commonwealth Games in 2002, now the home of Manchester City. The Olympic Park in the Lower Lea Valley will become a major regeneration once completed for the 2012 Olympics. Other grounds in regeneration areas: St Mary's, Southampton, Riverside, Middlesborough and Pride Park, Derby.

Urban regeneration

The figure below shows a model of urban regeneration in the heart of European cities.

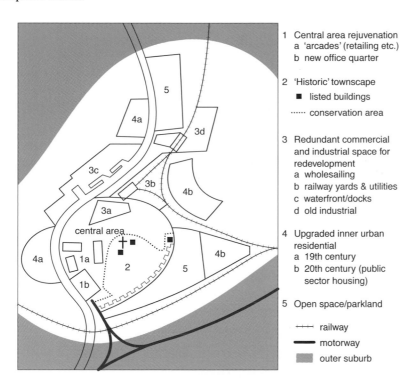

1 Central area rejuvenation
  a 'arcades' (retailing etc.)
  b new office quarter

2 'Historic' townscape
  ■ listed buildings
  ······ conservation area

3 Redundant commercial and industrial space for redevelopment
  a wholesailing
  b railway yards & utilities
  c waterfront/docks
  d old industrial

4 Upgraded inner urban residential
  a 19th century
  b 20th century (public sector housing)

5 Open space/parkland

  ┼┼┼┼ railway
  ▬▬ motorway
  ▨ outer suburb

For government regeneration schemes in the inner city, see pages 114–115.

Project tip

Projects about the issues surrounding the use of a brownfield site are manageable and can score highly if they are well researched.

Exam practice
answers: page 130

1 Why is regeneration of parts of MEDC cities necessary? (15 mins)

2 What are the consequences of reurbanisation on the economy, social make up and environment of urban areas? (25 mins)

Examiners' secrets

Use local examples in addition to those in textbooks. You will probably know them better than the examiner and score more marks if you have relevant detail.

The Central Business District

The CBD is the commercial and business centre of a town or city, which is often the original urban area. In some cases the CBD has migrated away from the old town, such as in Portsmouth and Southampton. It is the most dynamic and accessible area of the city and subject to constant pressure to change with the times. Large CBDs develop their own internal regions of specialisation.

Check the net

www.nationmaster.com/encyclopedia/
central-business-district,
http://www.london.gov.uk/gla/
publications/economy.jsp

Delimiting the CBD

The CBD may be a town's historic core. Studies in the USA have identified common features such as the high density of buildings, the height of buildings, and high land values. The CBD can be what people perceive it to be. Its boundary is often shifting, as areas become included (**zone of assimilation**) or discarded (**zone of discard**), which are part of the zone in transition. Parks, rivers and railway lines have often formed physical boundaries to CBD, as do major by-pass roads today.

How do geographers identify the CBD?

→ Mapping land uses – shops, offices, leisure, higher education.
→ Height index – business floors as a proportion of floors.
→ Intensity index – if over 50% of space is CBD.
→ Council tax per metre of street frontage.
→ An historic quarter that is conserved and adapted, e.g. solicitors' offices.
→ Pedestrianisation and/or traffic restrictions.

Structure of CBD

Retailing tends to cluster around the **peak footfall** area (sometimes seen as the **peak land value** point): this is where stores like Marks and Spencer are located. In malls there will often be **anchor stores** located at the ends of the walkways to pull shoppers along past other shops. Shops often cluster to assist **comparison shopping**. Quality of shops declines with distance from the core, with second-hand shops and money lenders/pawn shops on the fringe. Nearer the fringe there may be small offices and residences above the shops. In the largest cities, there may be segregation into areas of retail specialism such as the designer shops of the Golden Triangle in Milan, Bond Street in London and Faubourg St Honoré in Paris.

Financial retailing, such as Building Society offices and banks, tends to cluster close to retailing. **Clustering** can be measured by **Nearest Neighbour Analysis**.

Offices cluster and have been planned into separate districts. In larger cities the office areas will become more distinct and focused on key buildings. For example, in London, insurance clusters around the Lloyds Building, and banking around the Bank of England, with media firms (advertising, publishing) in the West End. Lawyers are often located in proximity to the Law Courts. Some cities will have a governmental office quarter (e.g. Cardiff), or even a local government area.

Links

For more revision on retailing, see pages 116–117

The jargon

Nearest neighbour statistic $R_N = 2D\sqrt{N/A}$ where D is the mean distance between neighbouring points, n the number of points divided by A the size of the area. R_N will range from 0 = totally clustered, to 2.15 = regularly spaced.

In some cities universities have become a part of the wider CBD, such as Oxford, Bristol and Portsmouth.

Recreation and leisure in the CBD is more than just hotels in the historic core. In the 20th century, recreational areas of bars and clubs were often developed in a separate area, catering for the lunchtime market from the offices but changing their character markedly after dark.

It will have transport termini on the fringe.

Challenges facing the districts of the CBD

→ **Office district**: expansion into redeveloped sites with higher building density; technology reducing demand for office space; newly emerging office districts such as La Defense, Paris and Canary Wharf, London; 'dead hearts' at weekends; the growth of science parks in the rural fringe.

→ **Retail streets**: see if you know what the threats are, and then turn to p108 to check your answer.

→ **Legal districts**: often take residential buildings of character such as Georgian houses, out of the housing market.

→ **Recreational areas**: threatening streets at night; vice districts.

→ **Access** from sprawling suburbs many miles away. Added to which might be congestion.

→ **Planning** has encouraged out-of-town development and companies find out-of-town cheaper and more attractive to investors.

→ **Conserving heritage** as a means of attracting shoppers and tourists e.g. Oxford, Bath, Chester, York. This can deter some developments.

Grade booster

When discussing challenges, be prepared to have both local and well-known examples to support your point. Be aware of the political dimension of the challenges because different parties in city and national government may see issues from different perspectives.

Other central area issues

→ Derelict retailing as new covered centres emerge.
→ Conversion of offices to gentrified apartments.
→ Homelessness in e.g. doorways; cars and public transport.
→ New tram routes and the environmental impacts (e.g. Sheffield or Croydon).
→ Park-and-ride schemes (e.g. Exeter or Winchester).
→ 'Empty hearts' to cities at night as people decant to the suburbs can be an issue.

Checkpoint

List some of the changes that you have noticed in a CBD that you have studied.

Action point

Can you draw a simple sketch map of a CBD, showing its main component areas and perhaps annotating it with the issues and challenges being faced in each area that you have identified?

Examiner's secrets

Do not use phrases such as 'as in Coventry' or 'this can be seen in Southampton' when you need to be more specific about the location of an activity within the CBD. Street or mall names imply detailed knowledge and understanding.

Exam practice   answer: page 130

For any named urban area, discuss some of the challenges being faced by planners and urban managers in its CBD. (20 mins)

Social structures and diversity

Social geography studies different ways by which social groups are recognised, the behaviour of different social groups (in buying a house for example), social processes such as **segregation**, and the distribution of different social groups. Housing is one of the main uses of space in cities and social geography focuses on the issues concerned with access to housing. **Social diversity** occurs as a result of the interplay of social groups and the increased ethnic and cultural background of urban residents.

Bid rent theory

→ Rent or land value declines with distance from the city centre.
→ The values placed on each area by each land use decline at a different rate.

Certain land uses dominate the bidding in different areas. The rent pattern is also affected by transport, local government policies and planning to give a complex pattern.

Housing

Housing choice is only there for those who have the money to make a choice. Choice is based on (i) life-cycle, (ii) social class and (iii) life changes. Therefore there is a pattern of choice for people in the UK over time.

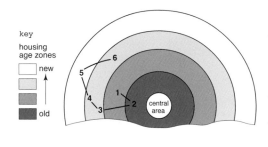

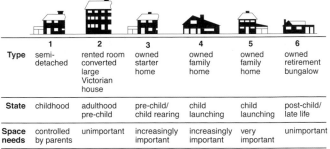

	1	2	3	4	5	6
Type	semi-detached	rented room converted large Victorian house	owned starter home	owned family home	owned family home	owned retirement bungalow
State	childhood	adulthood pre-child	pre-child/ child rearing	child launching	child launching	post-child/ late life
Space needs	controlled by parents	unimportant	increasingly important	increasingly important	very important	unimportant

key
housing age zones
□ new
↑
■ old

Housing is filtered downwards and upwards through social groups. This results in low-income, inner-city residents (the underclass) trapped in an area with social problems. Low-income residents can be trapped in peripheral social housing. In some areas the process is reversed as professional people move into an area – **gentrification**.

Choice constraints – these are negative externalities

→ Limitations may depend on the ability to pay.
→ There are markets for different types of housing.
→ There are fashions that increase desirability.
→ There are **gatekeepers** who approve finances for housing.

Encouraging choice – these are positive externalities

→ Areas approved due to their future potential or being the 'right address'.
→ Some areas become gentrified.
→ Physical attributes such as relief or on water/river front.
→ Near to services (e.g. restaurants, schools) which tends to then have a multiplier effect. Perception of areas from media and own experience.

Multiple deprivation

Multiple deprivation as a cycle of decline

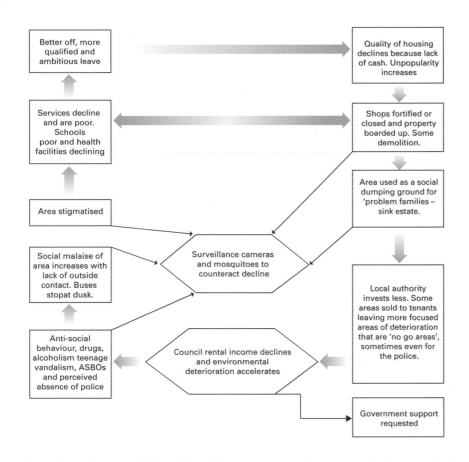

Ghettos as segregation

Why do people cluster?

→ Avoidance – focus on the community and its religious and cultural needs.
→ Preservation of the culture (language and religion) in home and neighbourhood.
→ Defence to help new migrants and give security.
→ Attack because together when threatened – riots.

How do ghettos grow?

→ Spillover – gradual outward spread, often towards perceived 'better' areas.
→ Leapfrogging to new areas.
→ Response to local policies – urban renovation pushing people out.

Studentification

The growth of areas of student population is especially marked in cities with more than one higher education institution. Cheap, rental housing in declining areas; although some cities do have purpose-built accommodation for students either provided by the private sector or a university.

> **The Jargon**
>
> *Gentrification* is a form of segregation where an area located close to a city centre is increasingly occupied by high earners and/or the movers and shakers of society (politicians and the media), and is often associated with reurbanisation, e.g. Islington. In the suburbs they might be gated communities.

Exam practice
answer: page 130

Account for the presence of socially segregated areas in cities. (20 mins)

Inner city issues

The inner city provides geographers with a range of issues to be investigated. Data sources are wide and varied, ranging from your own primary observations to secondary data from censuses, surveys and planning documents. This section focuses on political initiatives which are a response to the points raised in the sections on regeneration, diversity and deprivation.

Political factors

In most cases involving change in the inner city, political factors loom large. Decisions may be made in accordance with political ideology, in response to pressure groups, to satisfy a forceful, rich developer, or a big corporation. In all cases involving change it is useful to ask, 'Who gains what, where, how and why?' and conversely, 'Who loses what, where, how and why?' These questions are the substance of welfare geography.

Political initiatives

→ **Enterprise Zones**, 1981–96
  12 Urban Development Corporations, 1981–98
→ **City Grant and English Partnerships** (to fund private sector in priority urban areas; merged with Commission for New Towns)
→ **Single Regeneration Budget (SRB) Challenge Fund** 1994 (resources for up to seven years for economic, social and physical regeneration, fifth round 1999)
→ **New Deal for Communities (NDA)** one of the main funds in 2008. This aims to "tackle multiple deprivation in the most deprived neighbourhoods in the country, giving some of our poorest communities the resources to tackle their problems in an intensive and coordinated way." The partnerships tackle five key themes: poor job prospects, high levels of crime, educational under-achievement, poor health, problems with housing and the physical environment. Note that some of these tasks are relevant to other areas such as crime and health.
→ **Neighbourhood Renewal Fund 2001–8**, one of the main funds has been made available to 86 of England's most deprived local authorities, to enable them in collaboration with their Local Strategic Partnership (LSP), to improve services, narrowing the gap between deprived areas and the rest of the country.
  → UK's **City Challenge**
  → Programmes to recycle buildings – place hotels in old chapels etc, often using Urban Regeneration Companies
  → Better use of redundant floors, e.g. above shops

The **Regional Development Agencies** (RDAs) have taken over much of the responsibility and funding for regeneration. The distribution of funds is shown in the table. The effectiveness of regeneration schemes has yet to be proved, as there are doubts over how much impact they have, particularly on the difficult area of economic and social renewal.

RDA funds

First allocation City/town (district)	£m	Second allocation	£m
Bradford (Little Horton)	49.9	Birmingham (Aston)	54.0
Bristol (Barton Hill)	49.9	Brent (South Kilburn)	50.1
Leicester (Braunstone)	49.5	Coventry (Wood End, Henley Green, Manor Farm)	54.0
Middlesbrough (West)	52.1	Derby (Derwent)	42.0
Newham (West Ham & Plaistow)	54.5	Doncaster (Doncaster Central)	52.0
Sandwell (Greets Green)	56.0	Hammersmith & Fulham (North Fulham)	44.2
Southwark (Aylesbury Estate)	56.2	Haringey (Seven Sisters)	50.1
Manchester (Beswick & Openshaw)	51.7	Hartlepool (West Central Hartlepool)	53.8
Hackney (Shoreditch)	59.4	Islington (Finsbury)	52.9
Brighton (East Brighton)	47.2	Lambeth (Clapham Park)	56.0
Hull (Preston Road)	55.9	Lewisham (New Cross Gate)	45.0
Liverpool (Kensington)	61.9	Luton (Marsh Farm)	48.8
Newcastle (West Gate)	54.9	Oldham (Hathershaw and Fitton Hill)	53.5
Norwich (North Earlham & Marlpit)	35.2	Plymouth (Devonport)	48.7
Nottingham (Radford)	55.1	Rochdale (Old Heywood)	52.0
Tower Hamlets (Ocean Estate)	56.6	Salford (Charlestown and Lower Kersal)	53.0
Birmingham (Kings Norton)	50.0	Sheffield (Burngreave)	52.0
Total Round 1	**£895,379,176**	Southampton (Thornhill)	48.70
		Sunderland (East End and Hendon)	53.9
		Walsall (Bloxwich East and Leamore)	52.0
		Wolverhampton (All Saints and Blakenhall)	53.5
		Knowsley (North Huyton)	55.8
		Total Round 2	**£1,126,131,000**

Project tip

All these initiatives could be the subject of a good local project.

Check the net

http://www.regenerationuk.gov.uk/ has several case studies. See http://geographyfieldwork.com/ urbanregenerationissuesUKcities.htm for a discussion of Salford. http://www.neighbourhood.gov.uk/ page.asp?id=613 lists Neighbourhood Renewal areas. See also Making it Happen: Urban Renaissance and Prosperity in our Core Cities – A Tale of 8 Cities, 2004.

Other examples of initiatives

Sheffield Advanced Manufacturing Park: the University of Sheffield and the Boeing Company have set up an Aerospace Manufacturing Research Centre in the Park. This is attracting other research-based businesses to locate there. East Manchester was the centre of the Commonwealth Games 2002 and has a legacy of the new City of Manchester Stadium, and a velodrome. Manchester's city centre population has risen from under 1,000 in 1991 to over 15,000 today. Liverpool's has increased from 2,300 to more than 9,000 over a similar period and this should increase further with its 2008 European **Capital of Culture** status. Glasgow benefitted from similar status in 1990 when it was called City of Culture.

Exam practice   answers: pages 130–131

Consider an inner city area in the UK before and after regeneration.

(a) Describe the ways in which housing, streets and the general environment have been improved.

(b) Outline some ways in which you could use primary fieldwork and secondary research to study improvements resulting from regeneration in an urban area. (20 mins)

Retailing and services

The key theme in retailing is the tension between retailing in the city centre and new retailing both on the fringe and on regenerated sites in the suburbs. Online retailing is also a threat. Both offices and retailing are decentralising. The term tertiary sector is often used in the context of services. It comprises wholesaling, retailing, transport, the professions, leisure, public administration and personal service. It is based on goods. The quaternary sector is the sector that has been added by dividing up the service sector. It provides high levels of skill and education, such as research and development, financial management and administration. It is information/knowledge based.

Grade booster

To gain good marks when discussing retailing you need to demonstrate an awareness of the range of changes and not just focus on new out-of-town developments which demonstrate limited understanding of the whole.

Retailing

Checkpoint 1

For any one of the retailing types listed here, suggest reasons for its development. What might be the case raised by objectors to such a development?

Key

⌒ City centre Pedestrianised & Traditional – Kingston Guildford High Street Cascades, Portsmouth, Arndale, Manchester, Eton Square Newcastle

⌒ Malls

▭ Suburban Centres, Leytonstone High Road

◎ Brownfield Redeveloped Sites – Gateshead Metro Centre

· Corner Shops – mainly inner city

☐ Convenience Shops/Petrol Retailing

○ Early Retail Parks – Brent Cross

○ Retail Parks – Blue Waters, Merry Hill

⊗ Village/Small Town – Antiques – Arundel

S Supermarkets & Superstores – larger the further out

XXX Automobile and Fast Food Rows

═ Roads

═ Motorways

F Factory Outlets – Gunwharf, Portsmouth; Whiteley, Hants & Clark's, Somerset

Check the net

Road plans are very good sources of information on the layout of shopping centres. There are plans for the past available. There is data on all the major malls in the UK at www.ukwebstart.com One out of town site is www.bluewater.co.uk

Recent changes in retailing that you should know about are:

→ specialist shops returning to central area – delicatessens
→ supermarket flight to superstores in regional malls (US term)
→ arcades and covered city centre malls with specialist outlets – mobile phones, bookstores, cybercafés
→ underground retailing where climate can be harsh, e.g. Montreal and Toronto
→ superstores and retail parks on periphery, e.g. Bluewater
→ regional malls as a part of regeneration, e.g. Trafford Centre
→ motor rows – the clustering of car sales franchises
→ retailing within petrol service stations, e.g. Marks and Spencer in Moto service areas on motorways
→ factory outlets (Clarks Village, Somerset) and outlet malls (Gunwharf Quays, Portsmouth).

Threats to retailing in CBD

1) The increase in financial pseudo-retailing: banks are able to afford peak rents and outbid shops for space on the main street.
2) The 'sameness' of High Streets dominated by big companies, sometimes called 'clone towns'.
3) Pedestrianisation was an issue 20 years ago, although access for the disabled in such areas is still a challenge.
4) New covered shopping areas (such as West Quay, Southampton) have had a deleterious impact on the traditional High Street.
5) Access for deliveries.
6) New retail park developments on fringe or out of town.
7) Street markets/farmers' markets as a pull.
8) 'Tesco towns' where a retailing activity is dominated by one major company.

Not every retail development is successful. There are 'dead malls' in some city centres and even the outer suburbs, and dead areas often in old suburban centres.

Action point

Make your own notes explaining why a particular out-of-town retail development known to you was located in that place, and what the effects were on retailing elsewhere. Do you have examples of decentralisation?

Offices and service industries

The range of locations of offices in West European cities is shown below.

Checkpoint 2

Can you give examples of clusters? Why do some shops cluster and not others?

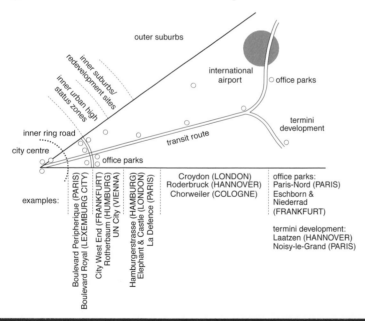

Action point

Suggest where each of the services listed might be located. What factors might influence the location of each service?

Types of services located in cities

Financial services	banking, insurance, stock market, currency trading
Business services	facilities management, copy shops, computer consultants
Public administration	government offices, police, local government
Trade services	hotels and restaurants
Education services	universities, colleges, schools, training organisations
Recreational and cultural services	concert halls, museums, stadia used for conferences
Health services	hospitals, vets
Communication and transport services	couriers, buses, travel agents, taxis
Personal services	nursing homes, day care, child care nurseries
Charities	national organisations, local outlets

Project tip

An analysis of any of the functions listed in a city or large town could become the subject of a project. This is a good idea if you are studying economics or business studies.

Action point

Draw and retain for revision, a sketch map to show the location of service activities in a city that you know. Why are they located where they are?

117

The urban fringe

The urban fringe of cities is the zone that benefits from accessibility to the urban area. Therefore its functions are linked to those in the urban area. It is a zone where conflicts occur between various interest groups.

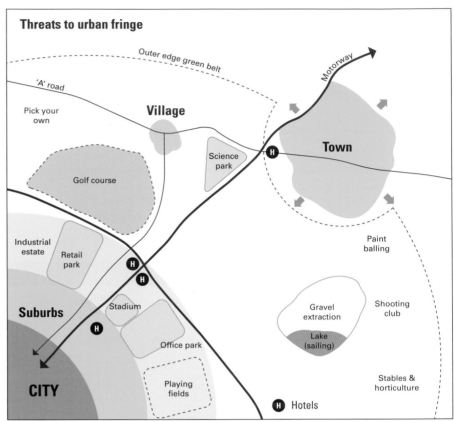

Developments in the fringe

→ Intensive and/or agribusiness such as market gardening and pick-your-own farms.

→ Playing fields that are used mainly by the urban dwellers for formal and informal (dog walking) recreation.

→ Major sports stadia (e.g. Kassam, Oxford) can be used for concerts.

→ Golf courses, shooting ranges and other extensive recreational sites.

→ Mineral extraction – sand and gravel – leaves derelict land until rehabilitated (e.g. as water parks) or reclaimed by landfill. Rolls Royce's factory is in Chichester in a disused gravel pit.

→ Retail parks at points with good access to a large catchment population. Occasionally built on landfill (e.g. Sainsbury's in Chichester). Can include auto parks – areas of car sales showrooms.

→ Office parks with good access for workforce from suburbia and rural areas.

→ Industrial parks – one of the original threats to the fringe – now replaced by Science/Technology Parks such as Cribbs Causeway in Bristol, and Cambridge Science Park.

→ Multiplex complexes.

→ Park and Ride termini (e.g. Oxford, Winchester).

→ Hotels such as Travelodge and Holiday Inn Express.

Check the net

www.naturenet.net/status/greenbelt.html,
www.durham.gov.uk/landscape/usp.nsf/
pws/Landscape+Strategy+–+Broad+Issues
+–+The+Rural+Urban+Fringe

The jargon

Green belt refers to land that has been legally protected from development that surrounds an urban area. London's green belt was established in 1948.

Controlling pressures on the fringe

→ Green belt e.g. London; green wedges e.g. the Ruhr; green zones and strategic gaps where all urban development is banned.
→ Permitted uses in green belts include mineral extraction, country parks and playing fields.

Issues in the urban fringe

→ Retail parks and superstores – impact on traffic flows, the environment and on other retailing.
→ Office parks and Science/Technology Parks have an impact on employment, commuting patterns and other areas of the city.
→ Leisure centres and new parks on reclaimed land, e.g. former gravel pits. Could include golf courses, golf ranges, new stadia and cinemas.
→ Marinas on reclaimed land (Port Solent, Portsmouth), or in old docks (Maryport).
→ Loss of school playing fields.
→ By-passes – building, noise, new edge for built-up area.
→ Sale of council houses and impact on the environment.
→ Problems of intrusion onto agricultural land.
→ Villages in green belt become 'dormitories' for upper middle classes to the exclusion of those on lower income, as no development is permitted bar some minor in-filling.
→ Problems of landfill sites.
→ Issue of what to do with former airfields now engulfed by suburbia, e.g. Brooklands, Surrey and Lee on the Solent.

Why do issues arise?

→ Population growth.
→ Rising living standards.
→ Changing economy.
→ Political pressures to meet the demands listed above.
→ Economic gain for landowners whether they be companies or individuals.
→ Variations in wealth and life chances of the people affected.

New settlements

→ **New towns** are hardly new now, as many are more than 60 years old. However, there are issues to do with further growth and spread of this generally successful means of accommodating urban growth (e.g. Harlow, Essex). The largest are cities in all but name (e.g. Milton Keynes).
→ Private new towns – many have now been developed (e.g. Bradley Stoke near Bristol or Hampton near Peterborough).
→ New villages and major village expansions, many of which are built to comply with demands for housing in the affluent South East of England.

Exam practice answer: page 131

Outline the pressures on the urban fringe of an urban area that you have studied. (20 minutes)

Rural settlement changes

Action point

Does this type of village exist today? In what ways do people try to retain this idyll?

Rural settlements were once those associated with people living and working in the countryside. Because of changes in the agricultural economy, and counterurbanisation, there are now fewer differences between urban and rural settlements in MEDCs.

What is rural?

City	Green belt	Commuter villages	Small town	Villages and hamlets	Remote rural districts
	Desirable villages and hamlets High cost property Agribusiness	Counterurbanisation Expanding settlements Non-agricultural uses Good transport links		Agriculture and long distance commuting Threat of new settlements	Declining villages Second homes Loss of social identity

← Distance from city edge →

Revised indices	Characteristic	Change since Cloke developed index in 1971
Percentage of workers in primary sector – agriculture & forestry (not fishing, mining which are also primary)	Very low due to mechanisation and farm mergers	Of declining importance
Percentage population over 65	More rural areas do have more retired	Better to distinguish the local retired from the retired who have migrated to area
Percentage of population in working age groups (16–65)	Will vary depending on distance from employment, early retirement moves, proportion in education 16–21	Because of counterurbanisation more are working. More women working
Percentage population change among resident population	Tends to be rising as more migrate to countryside although most remote still declining	Density of second home population might reveal remoteness
Distance from urban areas	The greater the distance the more rural the area	Still similar
Percentage of working population employed outside of village/hamlet	Will be high in commuter, counterurbanised areas and lower further afield	Teleworking can alter this generalisation
Percentage resident for less than five years	Will pick up the latest movers but misses those who arrived in earlier years	Ignores second home population
Population density	Declines with distance but can be high in some settlements due to in-filling. Lower in remoter areas	Still a characteristic
Occupancy rate – percentage of population at >1.5 persons per room	Affluence and smaller family size has brought lower densities per room in most areas. It will generally be low because of affluent in-migrants	Not a good measure and needs replacing by a measure of service provision within a given physical or time distance
Household amenities (baths, hot water)	Are there newer measures, such as two-car ownership and mobile phone access?	Not such a good measure today due to overall affluence

Potential indices could examine the perceived scenic value of an area, proximity to recreational opportunities, distance/time to schools, shops, doctor's surgeries show that living in a rural area impacts differentially on social groups.

Commuter villages

Counterurbanisation reinforced the growth of commuter villages sometimes called **suburbanised villages**.

Why did these changes take place in the late twentieth century?

→ New towns and planned growth of small towns put more jobs within easy reach of the country.
→ Improvements in rail and road transport.
→ Branch plant growth.
→ Diseconomies of scale in cities.
→ Improvements in communication – teleworking using Internet.
→ Jobs in areas of scenic quality (e.g. Yorkshire Dales).
→ Retirement migration to the countryside.
→ Planning policies such as greenbelts making areas desirable.

What have been the consequences?

→ New social classes based on service sector work in villages.
→ Reduced availability of homes for rural workers.
→ Social housing and the lower-paid are clustered in some villages.
→ General stores close because people shop in superstores in cities.
→ Closure of Post Offices.
→ Services for the more affluent, e.g. antique shops.
→ Declining public transport unless subsidised.
→ Better access to jobs in several more distant locations.
→ Villages in green belts attract the more affluent and those whose children have 'flown the nest'.

These develop as a result of affluence, rural depopulation and farm abandonment (e.g. in Tuscany) and changes in agricultural practice (e.g. barn conversions). Second homes result in improved upkeep and employment for those who maintain and improve property. Consequently, house prices rise which provokes opposition, the age structure alters, and the settlement may be 'dead' out of season. Champex, in the Alps, has 150 residents yet there are homes for 3000. In this village, retailing remains (garage, butcher, baker and post office), together with ski and walking equipment shops and souvenir shops, plus six restaurants, three estate agents and even a furnishing shop.

Project tip

Good topics for projects include: testing a revised Cloke's index, investigations into changing functions of a village; a contrast between a village within a greenbelt and one beyond a greenbelt; migration into a village – does it conform to ideas about either migration or counterurbanisation; using censuses to look at depopulation of a village; a study of second homes in an area and their effect on the area.

Check the net

A good example of what can be done to maintain a remote settlement in a tourist area is www.plockton.com. More about Champex can be found at www. champex.info/champex. Also Google 'rural depopulation' to find many useful sites.

Examiner's secrets

Do not get too caught up in the sensational. For instance, when discussing the drawbacks to second homes don't just focus on the burning of second homes. Mention the issue but spend more time on the more important issues such as affordable homes for rural workers.

Checkpoint 1

Why do people leave the countryside in MEDCs?

Checkpoint 2

Rural depopulation: what are the causes and consequences of this?

Exam practice
answers: page 131

1 Examine the effects of counterurbanisation on the economic and social geography of rural settlements. (25 mins)
2 Outline how the scenic quality of the environment has influenced population change within one rural area. (10 minutes)

Leisure and recreation

Leisure time is mainly used for recreation. Even when we are tourists we often make use of recreational facilities, e.g. the London Eye. People at work can make use of leisure facilities, e.g. conference delegates going out for dinner. Many facilities for recreation and tourism are found in the leisure zones of the central area of a city. Parks and greenbelt land are leisure areas. Shopping can be leisure! Geographers need to understand how it has grown, where people go and why, and the effects of tourism on the environment, economy and society, whether in cities or rural areas.

Types of leisure

Classified by:

- → type of environment – coast, mountain, wilderness, heritage, city
- → type of interest – walking, skiing, mountain biking, bird watching
- → length of stay – hours, days, weeks, months
- → wealth – package, types of accommodation (from exclusive boutique hotels to hostels)
- → season – winter sun or snow, Lapland tours, Easter in Rome.

Alternative classifications are:

- → mass tourism to major resorts over 1–2 weeks – including skiing
- → weekend breaks, including city breaks
- → personally generated using the Internet
- → business-related
- → ecotourism
- → cultural and religious pilgrimages – Hadj to Mecca
- → sports tourism – Barmy Army for cricket, Olympics 2008 and 2012
- → touring
- → short versus long distance
- → niche tourism – whale watching
- → educational – including field trips
- → cruising – tropical, cultural, ecological and wilderness.

The jargon

Leisure refers to those activities that take place outside of the working day/week. The choice of activity is decided by the person concerned.

Recreation is leisure taking place from home. It can be active, e.g. sports and clubbing, or passive, e.g. visiting a cinema. It can be organised, such as a visit to a soccer match or a theatre, or informal, such as sitting in a park. It can be resource-based, e.g. a city park, or user-based, e.g. a dry ski slope.

Tourism is a commercial form of leisure that usually takes place away from the home base. It normally involves staying away for more than one night.

What has caused the growth in leisure?

- → Shorter working time.
- → More money to spend.
- → Improved transport technology – access to out-of-town cinemas.
- → Increased holiday allowances – more money for city and weekend breaks.
- → Importance of well-being – parks, playing fields, gyms.
- → Increasing domestication of leisure – pubs that look like home.
- → Technology, from videos to stadium rock and the Internet.
- → Growth of the new service class doing routine jobs in homes, so releasing time for leisure.
- → Marketing of venues.
- → Improved information via media.

The elements of leisure in cities

Point location-based	Area-based
Cultural facilities	*Physical characteristics*
Theatre	Historic streets
Concert halls	Buildings of note
Cinemas	Monuments and statues
Exhibition centres	Churches, mosques and cathedrals
Museums and art galleries	Parks and open spaces
	Water – canal and river fronts
Sports facilities	Harbours and marinas
Indoor arenas	
Stadia	*Social attractions*
Golf courses and	Liveliness – atmosphere
playing fields	Language
Amusement	Local customs and folklore
Casinos and bingo halls	Security
Night clubs	Suits age group
Festivals	
Education and self-improvement	
Health clubs	
Adult education centres	

Urban Parks

A 2002 Government report on urban parks saw people retreating from run-down and dangerous public parks and paying for private health clubs and indoor play centres. According to official estimates Britain has 27,000 urban parks, covering 143,000 hectares.

Hotels as leisure providers

The figure shows a model of the distribution of hotels and other accommodation in a city. Use your knowledge of your home town/city to see if it conforms to the model. What other leisure activities do the hotels have? Activities nearby are served by the hotels e.g. around Disneyland Paris.

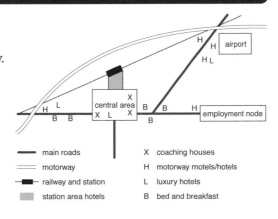

Decentralising leisure

Some leisure facilities have moved out from the central city: cinema complexes, new sports stadia (e.g. San Siro, Milan), concert venues, exhibition centres (e.g. NEC, Birmingham) and hotels.

Exam practice
answers: pages 131–132

1 Examine the factors that have assisted the growth of urban leisure. (15 mins)

2 With reference to one named urban area, discuss the impact of increased leisure facilities on the environment. (10 mins)

3 In what ways may tourist facilities also be recreational facilities? What planning problems arise from catering for both recreation and tourism in the same venues? (20 mins)

Deprivation

Deprivation can be applied to both people and areas. It expresses itself in employment, housing and services, which are linked to one another. Therefore, the term **multiple deprivation** is often used. Deprivation in both MEDCs and LEDCs can apply to both urban and rural areas. Deprivation is a set of interlocking cycles (see below).

The jargon

Socio-economic status is based on family background, education, values, income and occupation.

Check the net

www.un.org is an excellent starting point for information and data on settlement issues in LEDCs.

Action point

Can you apply the flow diagram to an area of multiple deprivation that you have examined?

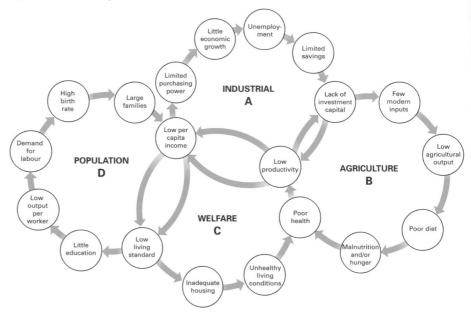

Rural deprivation in the UK

May be caused by:

→ **Remoteness** – distance from core area of a country, or hospitals, schools, police.
→ **Time distance** to reach facilities due to lack of public transport; distance will be greater for those without private transport.
→ **Social change** counterurbanisation and second homes – house prices rise due to external demand which puts housing beyond reach of local population. Elderly left in a rural area populated by younger affluent families and richer retirement migrants.
→ **Economic factors** – closure of shops due to competition from urban superstores and rationalisation of post offices, combined with distance, deprives the elderly and non-mobile of access to services.
→ **Loss of employment** opportunities in rural areas leading to unemployment of the rural unqualified.
→ **New rural underclass** (often run by gangmasters) who are cheaper to employ – often immigrants living in hostels and poor accommodation in nearby towns. In other areas this underclass may include the Traveller population (Roma and others).

Action point

Can you now list the causes of rural deprivation in LEDCs?

Action point

Consider the effects of changes in rural areas on the disabled.

Urban Deprivation

Urban deprivation varies between countries at different stages of development. Cities bring together people from different social groups, classes and cultures, some of whom are excluded and deprived of some of the benefits of urban life. The deprived, have-nots and the haves can live alongside one another and not interact, as indicated by the existence of favelas and shanty towns alongside affluent suburbs with gated developments.

What are the indicators of deprivation in an MEDC?

→ Low average income.
→ Poor health and eating habits.
→ Housing and Council Tax Benefits.
→ Housing quality (1960s tower blocks and council housing).
→ Educational achievement.

→ Unemployment.
→ Social instability – divorce, single parents, teenage pregnancy.
→ Child poverty and infant deaths.

Checkpoint

Can you describe a case of combating deprivation in an area of a town or city?

Multiply deprived Wards in inner Portsmouth, and a peripheral estate, Havant.

Indices	Charles Dickens Ward	Warren Park
Population 2001 Census	15,256	6951
No car (% of households)	64%	36.69%
Unemployed (Portsmouth 3.09%)	5%	4.8%
Aged 16-74 with no qualifications	41%	48%
Long term ill	24%	19%
Poor health	13.5%	11%
On state benefits	21.55%	ND
Child poverty rank UK	23rd worst	ND

Grade booster

Local area statistics and ward level data are very useful for supporting evidence for deprivation. You can round figures to make them easier to remember because the examiner will not know the same data. You will then demonstrate that you can apply your knowledge to the question which is a high level skill.

Deprived areas in LEDC cities

Spontaneous or **squatter** or **informal settlements** are defined by the UN as 'areas of housing units that have been constructed or erected on land to which the occupants do not have a legal claim'. They grow rapidly from migration and natural increase. According to the UN, in 2000, 837 million people worldwide lived in squatter settlements. Brazilian **favelas** house 36% of the population. In Addis Ababa (2.9 million) there are 2,000 hectares of squatter settlements housing 300,000. In Alexandria, 30%, and in Cairo 70% of the population live in illegal settlements.

How can governments help?

→ Provide a strong economy that is able to support house-building
→ Build new towns: Brazilia has built satellite towns
→ Provide site and service schemes.

Squatter settlements are part of a cycle: bridge headers establish settlement, become consolidators trying to improve the lot of the inhabitants and hope to be status seekers as they and the settlement are integrated into the urban area.

Examiner's secrets

Make sure that you have examples to back up points that you make. When discussing deprivation that you might have studied in the field, do have detailed examples, which might include street or even building names. Do not say 'as in Leeds'.

Exam practice   answer: page 132

Evaluate the attempts being made to improve areas of multiple deprivation. (30 mins)

Geography of crime

Geographers study crime because crimes have a spatial element, and there are links between the location of offences and the location of criminals. What constitutes a crime will vary between cultures, although you will focus on crime in the UK. People's perceptions of crime vary due to the influence of the media and politicians.

Check the net

www.cleanersafergreener.gov.uk,
www.ft.com/murdermap,
www.odpm.gov.uk.

Types of crime

It is possible to study the patterns within all of the following types of crime:

→ International crimes (e.g. drug trafficking, illegal migration and people smuggling).
→ Crimes against people (e.g. murder, assault, slavery, racial crimes and prostitution).
→ Crimes against property (e.g. burglary, animal theft, car crimes and graffiti).
→ Employment crimes (corporate crime, fraud and gang masters).
→ Youth crime and its gender variations.
→ Anti-social crime (e.g. vandalism, fly tipping, alcohol and drug induced crime).
→ Environmental crimes (such as fly tipping and pollution spills).

Checkpoint 1

Where does fly tipping take place in your area?

Checkpoint 2

Is there any pattern to casual rubbish dumping?

Location of crime

Geographers plot where crimes take place and explain why, whether inner urban, leafy suburb, rural, parkland, in the workplace. There can be 'go' and 'no-go' areas.

The time the activity took place may influence its location: recreation areas in cities are often the scene of night crimes against the person and property, as well as anti-social crimes. Crimes may be seasonal – stealing Christmas trees – or related to special events or sports fixtures.

The homes of criminals can be plotted from court records and press reports. Some offences take place within the same social area, whereas others involve locations further afield with the advantage of remoteness, lack of surveillance cameras, or an easy get-away.

People may be vulnerable to being a victim or an offender because of age, gender, ethnicity, household structure, occupation or lack of occupation, the social norms of a group, marginalisation, greed and poverty.

The characteristics of the physical and built environment may promote vulnerability to crime: street layout and lighting, surveillance cameras, building height and density, presence of vegetation and the amount of open space, and building design.

The impacts can be social, economic or environmental.

Murder in London

We can study homicide in several ways:

→ Its distribution – the map on the right shows data for 2007.
→ Its concentration. 13 wards out of 694 are the location for 10% of murders. (30% of all violent crime takes place on 1% of the streets.

→ Demographic characteristics such as age. The rise in killings of those aged 17 and under is a cause for concern.
→ Ethnicity – 51% of victims are from minority ethnic groups who make up 29% of the population.
→ Its association with the informal economy around drug markets.

Number of homicides by London borough, 2007

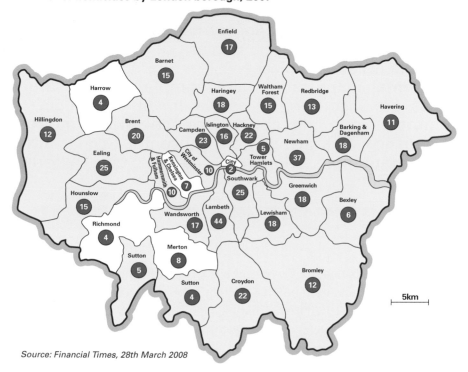

Source: Financial Times, 28th March 2008

Action point

Is crime concentrated in limited areas in your city? If so, why?

Project tip

If you have the opportunity to undertake group or personal work on crime it would be interesting to look at the distribution of surveillance cameras in a city. Why are they located where they are?

Combating crime

Measures include:
→ Safer and Stronger Communities Fund 2005. SSCF aims to tackle crime, anti-social behaviour and drugs, empowering communities, and improving streets and public spaces, in particular for disadvantaged neighbourhoods. In 2006, resources given to support initiatives in the most disadvantaged neighbourhoods, including 'Neighbourhood Element' and 'Cleaner Safer Greener'.
→ Local-scale initiatives such as Neighbourhood Watch, designing out crime, securing farm premises and equipment.
→ Private leisure spaces (the public now fear public spaces or regard them as so poorly maintained that they are willing to pay extra to enjoy the safe quality leisure time).
→ National scale initiatives e.g. CCTV, privatisation of public space including shopping centres and gated communities.
→ Drug and alcohol rehabilitation schemes – alcohol free zones.
→ ASBOs.

Exam practice   answer: page 132–133

How might different areas be more or less crime prone by day and night? (15 mins)

Examiner's secrets

Be careful when dealing with a sensitive issue such as crime data. If you use press reports, make sure that you carefully assess whether bias to a particular viewpoint is being displayed by the writer/presenter.

Perceiving environments

Cognitive mapping is the gathering of environmental knowledge via our senses' experience of the environment. Initially things are understood from an **egocentric** (self-centred) perspective. As our experience of the environment grows we develop a set of locations in relation to our home and an awareness of the paths between those locations. There is a type of map in our heads and it is based on direct experience, egocentrism, and locations and paths. We use this basic process of mental mapping when in unfamiliar places.

Mental maps seldom stay simple because larger areas are incorporated as we travel and begin to annotate our direct experience with indirect experience. This change from egocentric to **geocentric** understanding, orientates us to the external environment.

How do we see cities?

Kevin Lynch states that people see a city as made up of:

→ **paths** – our channels of movement along roads or rail lines or where we walk – they could even be what we remember from a sat nav device

→ **edges** – barriers that are not paths, such as a railway embankment or a motorway which cut one area off from another. Some edges are seams which bind an area together

→ **districts** – areas of a city identified by a common character, such as its architecture

→ **nodes** – focal points (often where transport meets or where there is a concentration of activity)

→ **landmarks** – distinct points with distinguishing features that people remember. These can be local such as Harrods' storefront, distant such as the London Eye or even outside the city such as Portsdown Hill, Portsmouth.

These all combine in different ways for different groups of people.

What factors influence the way people see an environment?

1. Age of individual
2. Gender
3. Experience and frequency of visiting that environment
4. Social class and value systems

Project tip

No matter where you live, an interesting way of finding out which parts of your area (or further afield) are most appreciated and valued is to ask people to write down, map or tell you where they would go for their last walk on earth.

Project tip

Get children to draw the map of their route to school. Analyse by gender, age and mapping skills. Why do their maps vary?

Boosting a city

Many cities and towns attempt to improve their image to attract industry, commerce and tourism. The process is called **boosterism** and is about creating a **capsule image** for the whole place. It is also applied to areas for tourist purposes (e.g. Guernsey, Herm, Alderney and Sark as 'the islands of choice', or Queenstown, New Zealand as 'the adventure capital of the world').

Some areas acquire a name that reflects current trends, such as 'Nappy Valley', the area between Clapham and Balham in London where many young professional families live. Humour often conveys how people sense places. The 'Costa Geriatica' sometimes refers to UK south coast retirement towns. Budleigh Salterton in Devon was labelled 'God's waiting room' by an eminent professor of education because of its very top-heavy population pyramid.

Estate agents, developers and local politicians often encourage people to imagine a false status for a neighbourhood (for example, former villages engulfed by suburbia continue to be referred to as 'the village' despite having several thousand inhabitants and being physically attached to the city). New Town neighbourhoods were often erroneously compared by developers to villages.

Project tip

Get children to draw the map of their route to school. Analyse by gender, age and mapping skills. Why do their maps vary?

Checkpoint

Could you summarise what makes a view good?

Grade booster

Psychology students should draw on their studies as it will ensure high marks if you relate your work to psychological theories and principals.

Perceptions of modern rural life

It is idyllic because	**It is not** because
Rich and poor live close together	Areas are often segregated
Agricultural workers live there	Few agricultural workers can afford to live there.
Everything is at hand in the village centre	Shops have closed or become antique shops
There is a thriving community	Many communities blighted by second homes
Social facilities such as pubs bring people together	Pubs becoming gastropubs and restaurants geared towards visitors.
Village school at centre of community	Educational rationalisation removing schools which are then sold for conversion to homes for affluent.

Exam practice   answer: page 133

Discuss the factors that influence an individual's perception of different areas of an urban area. (15 mins)

Answers
Settlement

Urbanisation

Checkpoint

A primate city is one which dominates the national urban system so much that it is exceptionally large and often several times larger than the second city. Paris and London are primate cities and at a smaller scale St Peter Port, Guernsey and St Helier, Jersey dominate the settlement systems of the Channel Islands.

Exam practice

1 Some are primate cities e.g. Tokyo is twice the size of Osaka. Other primates are Mexico City, Cairo, Seoul and Buenos Aires. Others are cities in the countries with the highest population numbers and some of the greatest densities of population in Asia in particular e.g. India and China. Only 6 are in the largest MEDCs. Some are the main centres in Latin America which were the cores of colonial and post-colonial development. The only African city is Cairo and there are none in Australasia because of its late urban development. Other than London, the main centres of urban population in the nineteenth century are all missing.

2 Similarities are: migration from rural areas, migration of the population seeking work, a flow of younger people into the urban areas, rapid natural growth of the population, poor housing conditions.
Differences in LEDCs are: the more rapid growth, generally poorer housing conditions – favelas and bustees, urbanisation without industrialisation focusing on cities as service industry centres (except possibly in modern China), greater urban extent, a phenomenon of the late twentieth century, whereas MEDCs were generally in the nineteenth and early twentieth century, and the transport system less developed.

Regeneration and reurbanisation

Exam practice

1 As city buildings and areas age, their layout, design and architecture are not easily adapted to more modern uses. In some parts such as nineteenth century inner suburbs, the buildings do not contain the facilities for modern urban life. Decline of industries due to deindustrialisation, the loss of rail yards, and the closure of old fashioned docks have given many cities the chance to regenerate large areas.

2 This is a question that demands some good case studies, preferably from more than one area being reurbanised. If you follow the structure of the question and select different urban areas for each part you will be on course for a high mark. Examples could be Wapping, Cardiff Bay, Gunwharf, Portsmouth etc. Economy: reurbanisation often involves the refurbishment, modernisation and conversion of buildings or the building of new blocks which employ people. Once completed, reurbanised areas attract a population that wants leisure facilities, services and infrastructure, all of which will employ people, be they fitness instructors, nannies or taxi drivers.
Social make up: these areas are often home to young affluent persons with higher than average incomes – such as the 'yummy mummies' of Nappy Valley (Clapham, London). Little social housing, unless scheme had to provide (e.g. for first-time buyers).
Environment: city parks and play areas, improved environmental quality, new street furniture and street art. Provision for cars – underground parking or problems of parking 4x4s.

The Central Business District

Checkpoint

The kinds of changes that you might see in many cities are: increased redevelopment for offices or covered shopping centres; types of retailing growing or declining; a local government quarter; a spread of offices; and changes in recreational districts.

Exam practice

Note that ONE city is required which should be upfront at the start. If you select more than one city the marker will only credit examples from the best developed. You do not have to cover all challenges but select maybe 3 or 4 and genuinely discuss the benefits and costs of these. It could be the decay of one part of the CBD, a new central covered shopping area, the loss of retailing to the out-of-town developments, leisure areas and the problem of night time leisure, pedestrianisation, traffic measures, access for disabled persons (to give you a few pointers!). In some urban areas there is the challenge for those who live in the CBD, whether they are the very affluent or the disadvantaged living on the streets.

Social structures and diversity

Exam practice

You could use one city for your answer, or indeed look at several because the question's content should be about the reasons for segregation. Carry a sketch map in your head for this type of question so that you can show that you know a place. You could cover segregation of income groups, ethnic groups, religious groups, students and even gated communities. Economic forces, housing types, size and age; the forces leading to ghettoisation; the need to cluster are all valid themes.

Inner city issues

Exam practice

(a) *Housing* by internal modernisation and extensions. *Streets* by removal of through traffic, traffic calming and partial pedestrianisation together with landscaping. Garages provided with back access routes. General *environment* improved by trees and ornamental beds,

wires placed underground. Some redevelopment,
e.g. slum clearance.

(b) The first part should give you some suggestions.
Primary and secondary data are defined on page
186. Primary work would involve mapping of the
improvements and measuring their impact through
environmental impact analysis and sample surveys
of residents to ascertain reactions to improvements.
Secondary work could involve finding out what it was
like before, from maps and photos. Census data could
give some idea of the changed population. All census
material is dated – currently 2001 census. Perhaps
there were alternative plans for the area that were not
used – another secondary source.

Retailing and services

Checkpoints

1 You should be able to answer this for several examples.
Consider why they have developed as they have and
what problems these developments might cause.

2 Comparison goods often cluster e.g. shoe shops.
Fashion retailers cluster in e.g. Bond Street. Cluster
happens because people often want to compare
products and make rational choices.

The urban fringe

Exam practice

Show that you really do know the area. A sketch map
might help to illustrate your points. Much will depend on
the place that you select. Pressures for housing, mineral
exploitation, recreation, by-passes, and out-of-town
retailing might feature in your answers. This particular
question does not ask for opinions but it could be that a
different question asks you to evaluate the pressures and
say which are most important.

Rural settlement changes

Checkpoint

1 Consider: lack of jobs in agriculture; lack of transport to
get to shops and servcies; because they're young and
wanting city life; unable to afford increasingly expensive
rural housing.

2 Depopulation is a form of population migration.
Its causes are:
• loss of jobs and mechanisation in agriculture
• poor housing conditions in rural areas
• land degradation due to over-cropping and drought
• the perceived lack of provision of rural services and
remoteness
• marriage leading to a move to a partner's home
• labour migration to the towns.
The consequences are:
• the abandonment of property and fallow land
• more second homes

• loss of services and public transport
• the loss of social cohesion in the rural community
• an ageing population because the young migrate
• more travelling to find goods and services in nearby
towns
• loss of male labour, especially in LEDCs.

Exam practice

1 It pays to give a definition of counterurbanisation. You
then have two parts to your answer which may overlap.
The *economic* geography might be that jobs are moved
to rural locations. Rural jobs are no longer dominant.
Besides that, people may be working from home –
teleworking. Income levels in rural settlements will be
higher and spending power more but that might only
be spent in towns. There may be more service work
such as house cleaning, gardeners and perhaps some
settlements may be able to support specialist shops.
Social geography might include the altering of the
social mix so that the settlements are more middle
class. There may be more children because those
who move normally have children already. Eventually
though it might become a village of older age groups
as children migrate back to the urban areas. The
social activities might change e.g. golf comes in and
the flower show is discontinued. Social and political
networking to keep the village as it was (i.e. 'what it was
like when I arrived') grows.

2 This is a typical short question. You must name your
area and make it realistic – not Wales but preferably,
the Brecon Beacons or the Dyffed Coast Park. Scenic
environment has an intrinsic value among purchasers
and sellers (agents will often include a view from the
garden). Sometimes it is the environment at close hand.
Generally those attracted are those who can afford the
premium for environment who will either be in the child
rearing stage of life or those whose kids have fled the
nest. Data would help for a specific settlement or two.

Leisure and recreation

Checkpoints

1 Hotels decentralise because it is easier for visitors
to gain access, there is more space to develop e.g.
conference facilities, many visitors are on business and
need places that are easy to find. Many businesses are
outside of the city and often on adjacent sites. Guests
have greater access to a wider area from outside the
city.

2 Much will depend on whether it is an MEDC or LEDC,
and whether you look at the countryside or coasts, at a
city or a green belt.
Environmental impact: trampling, path erosion
and vehicle tracks; airport and resort construction;
interference with longshore drift; rubbish left behind;
carrying capacity; loss of land; endangered species or
areas such as World Heritage Sites and SSIs. The theme
is *environmental sustainability*.

Economic impact: additional income; multiplier effect; employment multiplier; spending on travel, accommodation, food, souvenirs, local tours and car hire; impact on agriculture.

Social impact: employment; dangers of Westernisation of culture; illegal activities including sex tourism and drugs; migration out of rural areas to resorts.

Grade booster

Good answers on recreation will always rise above unsupported generalisations and refer to a wide range of examples, thus displaying both good knowledge and understanding.

Exam practice

1 Factors are: affluence of all age groups enabling new recreational venues to be established, shorter working week, socio-demographic changes such as later marriage and childbirth, planning of new zones such as central clubbing and bar area and out-of-town cinema complexes. Better transport – cars and ability to pay for taxis (wealth again). Existence of universities drawing more young to inner city. Fashions can result in more leisure, e.g. biking resulting in cycleways. Fitness leading to more sports venues. All these should be supported by examples from a range of places, with a focus on your home town and one other city.

2 Much will depend on your choice of urban area. This question can be about a part of a city – e.g. a segment of London. Environmental impact could be on the greenbelt or on parks. It could be the impact of traffic, parking, noise, litter and even light pollution from stadia. Consider the impact of artificial environments. Is there an impact on wildlife?

3 Definitions would be useful. They are both when the supply is of local, national and international importance, e.g. many of the sights of London or Paris. Sometimes they are planned as such, e.g. Covent Garden. They are used longer if they satisfy two demands. The planning problems are to do with access – tourist buses clogging up streets and spoiling the experience for more local visitors. Tourism requires hotels not too far away, which might destroy other local recreational amenities. Provision of cafés and restaurants, which have to cater for two types of demand. Overcrowding of honeypots. Note that this question does not actually say 'in cities'. Therefore, you could look at other cases, such as the seaside resort, which may be a day trip for some and a holiday venue for others. This is a question where you can use your project work.

Deprivation

Checkpoint

Combating deprivation in Bellenden, Peckham.

The scheme began in 1997 because the area had 33% means-tested benefit receivers and poor private housing (some without inside toilets). Southwark council involved the community in improving social and private housing

and rundown services; street furniture was improved (some designed by Antony Gormley); better lighting designed by Zandra Rhodes and more street art added (sometimes called *facadism*). A market was improved to reflect the culture of the large Caribbean minority population. The Council saw it as a success but people ask, "what did it do for us"?

The National Neighbourhood Framework encourages community involvement by encouraging residents to determine how their area is managed. Tenant management organisations combat vandalism by speeding up cleaning and repairs, erecting quality street lighting, using vandal- and graffiti-proof materials, widening footways and narrowing roads, creating safer play areas and removing abandoned vehicles.

Exam practice

This is an example of a question which might be asked at A2 level because it has a high level command 'evaluate'. Other high order commands are 'assess' and 'to what extent'. A2 questions are designed to 'stretch and challenge' your knowledge and writing skills. Therefore, you must make sure that your discussion of policies does actually do more than describe. You MUST conclude by saying which are the more successful in your opinion and why they are more successful. It is typical of geography that the policies that you might discuss are covered elsewhere in your revision. (see pages 114–115)

Geography of crime

Checkpoints

1 Often outside of official amenity dumps when closed. Garage areas of residential environments. Along verges in rural areas. It might be possible to see abandoned vehicles as fly tipping. Into streams.

2 Very often in association with fast food outlets. Thrown onto roadside verges at points where traffic slows. Stream courses in urban areas. The keys are the opportunity to dump waste and lack of facilities to encourage proper disposal. Where crowds gather such as outside stadia. Around fast food outlets and motorway service areas. Chewing gum could feature –look at pedestrianised areas!

Exam practice

Crimes normally associated with night might include: vandalism on public transport routes (rail bridge tagging); crimes against the person in dead areas such as parks and alleyways; possibly burglary; drink driving; drunkenness and assault in night entertainment districts; and activities in red light districts. Day crimes might include: burglary (much more likely when people at work), vandalism; minor motor crimes; robbery when shops are open; shoplifting. From this list (and other crimes that you select) you will need to say whether the location does change. Some crimes such as soliciting on the streets (kerb crawling is an offence) might be there all the time, whereas crimes in entertainment districts tend to be focused on evening and night time. You should show some knowledge of crime locations in a town or city that you have studied.

Perceiving environments

Checkpoint

Not an easy one to answer because it will depend on age, social class, gender, ethnicity, and what one knows about places. Grandeur, vistas, changing seasonal colours, a personal memory, solitude, silence, and action could all be valid reasons. Do you know why you selected your list?

Exam practice

Among the factors that you might use are: age, family status, social status, gender, pathways taken through the area, official and unofficial information, the media's reporting of events in an area. You would need to expand on each factor from your own researches or your reading. One useful way is to contrast your views with those of your parents and grandparents because that will probably cover the first four factors in the list above.

Revision checklist
Settlement

By the end of this chapter you should be able to:

1	Explain what is meant by an urban (settlement) hierarchy.	Confident	Not confident. **Revise** pages 106–107
2	State the factors that cause reurbanisation, and its effects.	Confident	Not confident. **Revise** pages 108–109
3	Draw a sketch map of your home town or city's CBD, noting the dominant functions of the areas within it.	Confident	Not confident. **Revise** pages 110–111
4	Identify definite functional clusters and explain why clusters exist.	Confident	Not confident. **Revise** pages 112–113
5	Explain the processes leading to areas becoming gentrified or leading to areas becoming segregated from the rest of the city.	Confident	Not confident. **Revise** pages 112–113
6	Be able to assess whether your home urban area is improving as politicians suggest.	Confident	Not confident. **Revise** pages 114–115
7	Explain the location of different types of retailing in a city.	Confident	Not confident. **Revise** pages 116–117
8	Describe the different city locations for retailing and services.	Confident	Not confident. **Revise** pages 116–117
9	Give examples from your local area of different issues on the urban fringe.	Confident	Not confident. **Revise** pages 118–119
10	Summarise the characteristics of a remote rural area.	Confident	Not confident. **Revise** pages 120–121
11	Explain how leisure is changing in response to technological change.	Confident	Not confident. **Revise** pages 122–123
12	List the causes of deprivation in contrasting areas such as town and country, or MEDCs and LEDCs.	Confident	Not confident. **Revise** pages 124–125
13	Explain the pattern of vandalism in an area.	Confident	Not confident. **Revise** pages 126–127
14	Draw your own mental map of your home area using the Lynch terminology, compare it with reality and explain why you included or neglected to include features.	Confident	Not confident. **Revise** pages 128–129

Economic change

In the 21st century, the study of economic geography has shifted to explaining the patterns of interdependence between producers and nations. No matter where you look the forces of globalisation are apparent. These forces are driven by the transnational companies whose activities are more important than those of governments. 'A transnational corporation is a firm which has the power to co-ordinate and control operations in more than one country, even if it does not own them.' (Dicken)

A consequence of globalisation is the increasing disparity between people and regions within countries, and between countries themselves in the form of the **development gap** (differences between the levels of economic and social development in different regions of the world). Your studies of development should include a critical look at how levels of development are measured and whether the measures accurately reflect the position of countries, and regions within countries.

Exam themes

- Changing patterns of world economic activity
- The role and impact of TNCs
- Global inequalities in economic and social development
- The 'Development Gap' as a product of economic development
- Bridging the Development Gap

Topic checklist

	Edexcel		AQA		OCR		WJEC	
	AS	A2	AS	A2	AS	A2	AS	A2
Globalisation		●		●				●
TNCs		●		●				●
Levels of development		●		●				●
Inequalities		●		●				●
The development gap		●		●				●

Globalisation

Check the net

Company websites might provide you with up-to-date information. A word of warning: remember that company sites are there to promote the company and will preach their own view of themselves and their activities – they are likely to be biased.

Globalisation is the changing geography of economic activity in the world, brought about by the growth of new activities, the patterns of investment of multinational companies, and government policies towards economic growth. It involves the shift of production and services to lower cost locations. The consequences may be deindustrialisation in MEDCs and industrialisation in NICs, RICs and LEDCs. Globalisation has environmental effects e.g. spreading the sources of CO_2.

Globalisation as a stage in the development of MEDCs

Five stages of economic development (or **Kondratieff Waves**):

K1: the early industrial revolution.

K2: the **Age of Steam Power**, post 1840s. Led to coalfield industrial regions, e.g. South Wales, and the Ruhr.

K3: the age of **electrical engineering**, post 1890, which led to market orientated around large cities.

K4: the introduction of mass production or **Fordism** in the 1930s e.g. the development of car assembly plants. This is still present.

K5: the period since the 1990s, when **IT and biotechnology** – together with flexible production, new communications technologies and globalisation – have brought about a new range of activities.

Checkpoint 1

Name some Fordist industries.

Checkpoint 2

Name a flexible production industry.

Newer activities replaced the declining K2 and K3 industries. New activities prefer new locations – old industrial buildings are abandoned because they are unsuitable for modern technologies. Governments ameliorate the effects of decline by offering Regional Development Aid.

Why globalisation?

→ Labour costs are lower in labour-intensive industries, e.g. cotton. Making clothing has shifted to low-cost countries, although some more expensive products are still made in MEDCs.

→ Mass production shifts to lower-cost countries. Computer assembly moved to China and South Korea, from Singapore and Taiwan. Outsourcing is cheaper e.g. call centres.

→ Firms expand to gain a direct market in different countries (e.g. cars). TNCs such as Mittal, Microsoft, Ford, Toyota, Sony need new markets for continued growth.

→ New financial trading activities (e.g. securities houses).

→ New financial trading activities to meet demand of investors.

→ **Tariff blocs** exclude imports in favour of home products.

→ New technologies enabling e.g. rapid money transfers.

→ The role of transport – containerisation.

→ Government deregulation.

Action point

Select one industry from textiles, car assembly and electronics. Draw maps to show its changing global pattern of production.

Globalisation and deindustrialisation

Deindustrialisation is the decline of regionally important industries:

→ Competition from overseas (e.g. the effects of cheaper ship building in Japan and South Korea on ship building in the UK).
→ Products being at the end of their life cycle and not in demand.
→ Labour management outmoded, e.g. job demarcation.
→ The need to rationalise production.
→ **Just-in-time** technologies remove the need for warehouses.
→ The removal of government subsidies for old industries.
→ Companies taking production to NICs, RICs and, more recently, to Eastern Europe.

The effects of the global shift of services on India

50% of the world's top 500 companies now **outsource** activities to India with an estimated $24 billion benefit to the economy in 2008. Main centres are Bangalore, Gurgaon near Delhi, Chennai and Hyderabad.

Estimated global migration of jobs in the financial services sector

Of the estimated 13m jobs in financial services in North America, the EU and Switzerland, and Asian developed economies, around 2m are forecast to have gone to Indian Ocean Rim countries by 2008.

The jargon

'*Fordism* is the development of assembly-line processes that permitted the production of large volumes of standardised products.' (Dicken 1998)
Just-in-time is the delivery of goods just when they are needed. This was developed by the Japanese as a means to reduce storage capacity.
Just-in-case is the delivery of goods and raw materials in such a way that they can be stored for use when they are needed. It is an extravagant use of space and has been replaced by just-in-time.

Checkpoint 3

What might be the costs and benefits to India and the UK of outsourcing?

Estimated global migration of jobs in the financial services sector

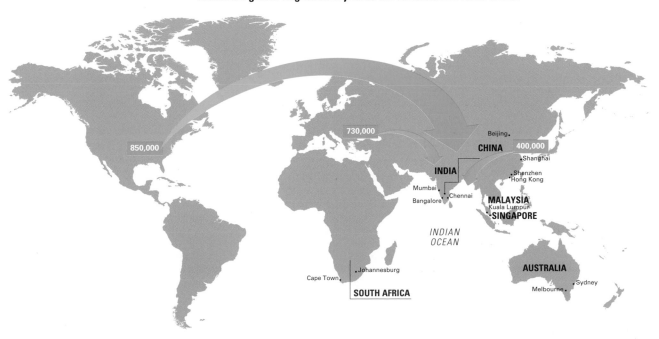

Estimated global migration of jobs in the financial services sector

Grade booster

High scoring students will note that economic globalisation is not necessarily matched by social gloablisation, which is slower to embed. Labour migrants are opposed, and cultural stereotypes are promoted, by the tabloid media.

Exam practice   answers: pages 146

1   Examine the relative importance of the factors giving rise to globalisation.
    (20 minutes)
2   (a) What are the causes of deindustrialisation? (10 mins)
    (b) Examine the impact of deindustrialisation on a region. (15 mins)

Transnational corporations

Transnational corporations (TNCs) are the major force in the global economy, although national governments and international organisations such as the World Bank are important. As the table below shows, they exist in all sectors of the modern economy and even in agriculture. TNCs invest in branch plants overseas, create networks of firms supplying the main company's product components, they collaborate with research organisations, and are financed by the international finance system and national governments keen to acquire their production and employment.

The top 20 global companies ranked by market value 2003–08

Rank 2008	Company	Country	Sector	Turnover $ m (rounded)	2003 Ranking
1	Exxon Mobil	Houston, USA	Oil & gas	452,505	*Microsoft*
2	Petro China	Beijing, China	Oil & gas	423,996	*GE*
3	General Electric (GE)	New York, USA	General industrial	369,569	*Exxon*
4	Gazprom	St Petersburg, Russia	Oil & Gas	299,764	*WalMart*
5	China Mobile	Hong Kong	Telecommunications	298,093	*Pfizer*
6	Ind. & Com. Bank of China	Beijing, China	Banking	277,235	*Citigroup*
7	Microsoft	Seattle, USA	IT software	264,132	*J & J*
8	AT&T	New York, USA	Telecommunications	231,168	*Shell*
9	Shell	London/Netherlands	Oil & gas	220,110	*BP*
10	Procter & Gamble (P&G)	Cincinnati, USA	Household products	215,640	*IBM*
11	WallMart	Louisville, USA	Retailing	265,906	*Americ Int.*
12	Petrobras	Rio de Janeiro, Brazil	Oil & gas producer	280,391	*Merck*
13	Berkshire Hathaway	Omaha, USA	Insurance	206,924	*Vodafone*
14	Nestlé	Vevey, Switzerland	Food producer	197,215	*P & G*
15	HSBC	London, UK	Banking	195,768	*Intel*
16	BP	London, UK	Oil & gas	191,844	*Glaxo*
17	Johnson & Johnson (J&J)	New Jersey, USA	Pharmaceuticals & biotechnology	183,751	*Novartis*
18	Total	Paris, France	Oil & gas	178,554	*Bank of America*
19	Chevron	San Francisco, USA	Oil & gas	177,265	*NTT*
20	China Const. Bank	Beijing, China	Banking	177,473	*Coca Cola*

Source: FT Global 500 2008 and 2003

Example of a TNC: the Volkswagen Group (VW)

VW is the world's fourth largest vehicle production company with 44 plants, 11 in Europe and seven worldwide. It was founded in 1938 in Germany. VW used seven strategies (six spatial and one technological) to become a Europe-based global manufacturer:

→ Growth in West Germany: Wolfsburg, Braunschweig, Hannover, Emden and Kassel.
→ Expansion on each continent: Mexico (1967), Brazil (1981), China (1982–85) and India (2000).
→ Expansion into Eastern Europe after 1989: Skoda, Czech Republic (1990) and Trabant in former East Germany (1990), Poland (1991).
→ Acquiring other companies in Germany (Audi), and opening plant in Hungary (1993).
→ Acquiring producers in Western Europe: SEAT (2002), Bentley and Lamborghini (Belgium and Portugal).
→ Developing luxury cars: Bentley (1998), Cosworth (1998).
→ Technological factors: Badge Engineering where different cars have the same components, reducing costs rather than developing globally identical cars. Parts are moved on a **just-in-time** basis and sourced locally, so that high cost assembly lines in Germany are supplied by lower cost parts from Poland, Hungary and Slovakia.

VW illustrates the following aspects of global location strategies:

→ The role of markets.
→ Political aspects of locational decisions.
→ Using government aid to assist establishing new plants.
→ The role of the global region of origin (Europe) in expansion.
→ The role of transport between manufacturing sites.
→ The importance of board room decisions that result in takeovers and mergers.

Checkpoint

What does the table tell you about changes in the location of control of TNCs since 2003?

Grade booster

Try to research your own multinational company rather than the one in the textbooks. Researched examples will score more than those learned by heart from textbooks.

Check the net

www.vw-personal.de/www/en.html is the site where you can find detail on VW. Visit www.barcap.com and click global reach to find a map of its offices worldwide.

Examiner's secrets

Have your own study of the distribution of a major manufacturing TNC that you have researched yourself, and that is different from the other students in your group.

Exam practice   answer: pages 146

What are the benefits and costs brought by TNCs to the countries in which they operate? (20 mins)

Levels of development

> *"Human poverty is more than income poverty – it is the denial of choices and opportunities for living a tolerable life."*
>
> UNDP Human Development Report 1997

You need to know how development is measured, why it varies and what might be done to alter the patterns of development. Development has three dimensions, economic, political, and social, whose impacts can be on the economy, government and society as well as the environment. There are different attitudes to development in different parts of the world. Levels of development can vary within a country as well as between countries.

The major economic groupings

Check the net

www.worldbank.org, www.oecd.org, www.imf.org, www.wto.org, and www.unctad.org

The threefold world of MEDCs, LEDCs and NICs is simplistic. Most geographers and economists recognise that there is a continuum of levels of development, and that the groupings are just a convenience for labelling.

The major groups of economies

	G8		MEDC		NIC/RIC		Oil rich		Former Soviet		LEDC		LDC	
	USA	UK	NL	Sp	Sing	SKor	SA	UAE	Pl	CzR	Bd	Gh	Bdi	Eth
GNI	43740	37600	36620	25360	27490	15830	11770	23771	7110	10710	470	450	100	160
GNP/cap	29240	21410	24780	14100	30170	8600	6910	17870	3910	5150	350	390	140	100
Urbanisation	77%	90%	90%	78%	100%	64%	87%	90%	63%	75%	26%	36%	9%	16%
HDI rank 2005	12	16	9	13	25	26	61	48	35	32	140	135	167	169
	0.95	0.94	0.95	0.95	0.92	0.92	0.61	0.81	0.83	0.9	0.55	0.55	70.4	90.4
GDI 2001	0.94	0.93	0.93	0.91	0.88	0.87	0.74	0.80	0.84	0.85		0.56	0.33	

GNI = **Gross National Income** the sum added by all producers, plus taxes and income from abroad in $US.

HDI = **Human Development Index** measures average achievement in three basic dimensions: a long and healthy life, knowledge and a decent standard of living, ranking from best to worst.

GDI = **Gender-related Development Index** measures the three dimensions of the HDI adjusted to account for inequalities between men and women. This reduces the values.

NL = Netherlands, Sp = Spain, Sing = Singapore, S Kor = South Korea, SA = Saudi Arabia, UAE = United Arab Emirates, Pl = Poland, CzR = Czech Republic, Bd = Bangladesh, Gh = Ghana, Bdi = Burundi, Eth = Ethiopia.

LDCs or Least Developed Countries have been defined by the UN.

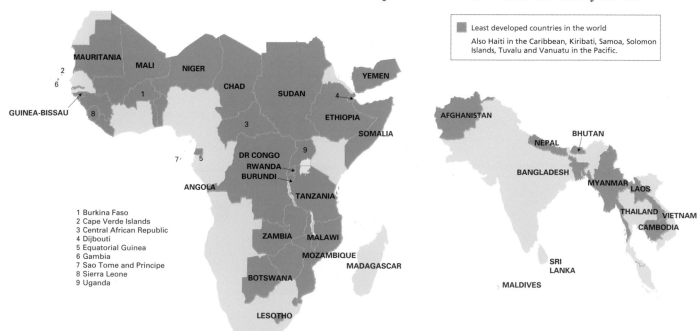

Least developed countries in the world

Also Haiti in the Caribbean, Kiribati, Samoa, Solomon Islands, Tuvalu and Vanuatu in the Pacific.

1 Burkina Faso
2 Cape Verde Islands
3 Central African Republic
4 Djibouti
5 Equatorial Guinea
6 Gambia
7 Sao Tome and Principe
8 Sierra Leone
9 Uganda

UN Human Development Index 2005 – selected countries

HDI rank	Country	HDI	GDP per capita $	Life expectancy at birth	Literacy rate
1	Iceland	0.968	36510	81.5	99%
2	Norway	0.968	41420	79.8	99%
4	Canada	0.961	33375	80.3	99%
12	USA	0.951	41890	77.9	99%
16	UK	0.946	33238	79.0	99%
25	Singapore	0.922	29663	79.4	92.5%
70	Brazil	0.800	8402	71.7	88.6%
81	China	0.777	6757	72.5	90.9%
112	Egypt	0.708	4337	70.7	71.4%
128	India	0.642	3452	63.7	87.0%
140	Bangladesh	0.547	2053	63.1	47.5%
148	Kenya	0.521	1240	52.1	73.6%
161	Rwanda	0.452	1206	45.2	64.9%
176	Burkina Faso	0.370	1213	51.4	23.6%
177	Sierra Leone	0.336	806	41.8	34.8%

Checkpoint 1

'Brandt's North and South is a gross oversimplification.' Show how this is so.

Action point

What is dependency theory? Make your own notes on this theory.

Other measures of development

Gross Domestic Product (GDP) per capita: total value of goods and services produced by an economy in a year, per capita.

Gross National Product (GNP) per capita: total value of economic production of a country divided by its population. It is a standard measure of economic development.

Human Poverty Index (HPI): longevity, knowledge and a decent standard of life.

Physical Quality of Life Index (PQLI): basic literacy rate, infant mortality and life expectancy at age one.

Health measures such as **infant mortality** (deaths of children under one year old per thousand live births), **patients per doctor**, **life expectancy**, and **birth rates**.

Calories per day: food intake.

Social indicators: literacy levels by gender, students at various levels of education, students per teacher.

The jargon

HDI was first used by the UN in 1990 and uses income per capita, adult literacy and life expectancy, equally weighted. The interval between the highest global value for each variable and the lowest is 1 and a country's actual value falls between 0 and 1. It recognises that development is not just economic.

Examiner's secrets

Questions often use the word 'evaluate'. This is a high-level skill. You must be able to say what the good and bad points are about an idea and critically compare them with other ideas or solutions.

Examiner's secrets

Watch for outdated opinions. Roman Catholic European countries have some of the lowest birth rates and not the highest among MEDCs. Even in LEDCs the rate is not the highest.

Exam practice   answer: pages 146–147

Critically evaluate the various criteria used to measure disparity between countries and within countries. (25 mins)

Inequalities

The jargon

The *Inequality Ratio* is used by the World Bank and is the ratio of income or consumption shares of the richest 20% to the poorest 20% of the population. The higher the ratio the more unequal the income distribution.

Global inequalities are a product of history as well as current political, economic and social decisions. The processes of development have affected countries in different ways, and have often increased inequalities within countries.

Economic models of inequality and disparity

1. Friedmann's model of development is often called the **core-periphery model**, and is shown below:

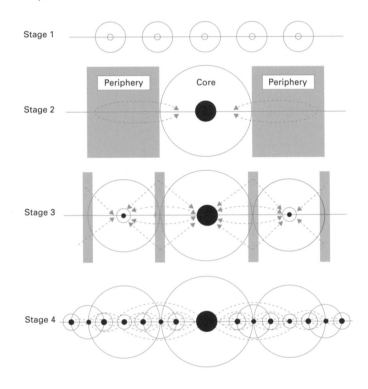

2. Myrdal developed the **cumulative causation** model.
3. Frank's **dependency theory** draws attention to the inter-connectedness between dominant MEDC trade and the periphery dependent for investment.
4. Nurske's **vicious cycle of poverty theory** shows the LEDCs trapped in a downward spiral caused by lack of capital and low incomes.

The social causes of inequality

→ **Demographic issues linked to poverty**:
   → high birth rates (Yemen, Ethiopia 49/000, Italy and Spain 10)
   → infant mortality rates (Uganda 120/100 live births, Japan 4)
→ **Food and malnutrition issues**:
   → food import dependency (Ethiopia 15%, Sierra Leone 21%)
   → malnutrition – 800 m are malnourished (37% in Africa, 20% in Asia)
→ **Health issues**:
   → access to safe water (Ethiopia 25%, Myanmar 38%)
   → access to doctor or nurse, especially at birth
   → disease control – vaccines, new diseases, AIDS

→ **Education opportunity issues**:
   → illiteracy – higher among females (adult illiteracy India F62%, M35%)

- → enrolled in primary school – lower for females (Ethiopia F19%, M27%)
- → **Childhood exploitation issues**:
  - → sexual exploitation of children – sex tourism
  - → child labour – 75 million aged 8–15 working to pay off family debts (16 million in India and 7 million in Brazil)
- → **Gender issues**:
  - → access to contraceptives – links to demographic issues above
  - → deaths in pregnancy (Africa has 20% of world births and 40% of deaths in pregnancy and childbirth; Europe has 11% and 1%)
  - → genital mutilation (100 million people) – leads to infections in up to 80% of cases
  - → women not represented in higher-level jobs; tend to be in lower-paid and less skilled occupations (in LEDCs can be tradition too, e.g. work in fields).

The links below summarise the inequality issue for women in LDCs.

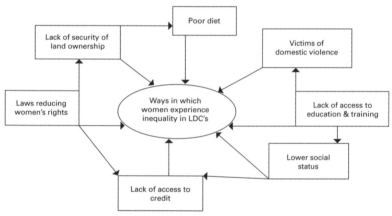

Political factors

- → **Colonialism** favoured the initial development of port cities in LEDCs. It also led to the growth of strong companies exploiting the colony but based in the colonial power e.g. East India Company.
- → Marx's political model of development saw the world developing from feudalism through capitalism and socialism to communism, in contrast to Rostow's evolutionary capitalist model.
- → Debt burdens of LEDCs who borrowed heavily in the 1970s and are now unable to repay the debts.
- → Political instability with military coups and repression that cause investors to hold back and withdraw investments.

Environmental factors

- → The presence of harsh environments such as deserts and areas where the climate is unpredictable, such as the Sahel.
- → Lack of water for irrigation, salinisation of water supplies and soils.

Checkpoint

Countries with high income inequalities	Inequality ratio
Sierra Leone	58
Namibia	56
Lesotho	44
Bolivia	42
Central African Republic	33
Botswana	31
Paraguay	28
Haiti	27
Columbia	25
Panama	24

Comment on the distribution of the countries with high income inequalities shown in the table.

The jargon

Neo-colonialism is used to explain how colonial influence continued after most countries achieved independence from their colonial masters last century. It refers to the continuing influence of the former trade links on the economic development of a country.

Check the net

www.worldbank.org/wdr and also
www.worldbank.org/data
www.hdr.undp.org

Exam practice   answer: page 147

Examine the relative merits of selected theories of unequal development.
(25 mins)

The development gap

One of the strengths of geography is that it looks forward to offer solutions to improve the world as the home of its inhabitants. Prediction is a dangerous activity because it can be wrong when looked back on 20 years later. Nevertheless, a good geographer should be prepared to look at the potential solutions to issues, and evaluate them.

Evidence for the development gap

→ Levels of debt.
→ Migration, economic and illegal migration and brain drains.
→ The health of a population – including dependency, impact of diseases and pandemics.
→ Population and resource balance – food insecurity.
→ The impact of conflicts, e.g. Democratic Republic of Congo and Darfur.
→ All of the indicators used on the previous pages of this guide.

Addressing the development gap

Millennium Development Goals (UN 2000)

Goal	Target	Indicator
1. Eradicate extreme poverty and hunger	50% reduction child malnutrition 1990 to 2015	% earning under $1 a day, hunger among children under five
2. Achieve universal primary education	Achieve by 2015	% children completing last grade of primary school
3. Promote gender equality & empower women	Achieve gender equality in school enrolment by 2005	Ratio of boys to girls in school enrolments Women employed in non-agricultural sectors
4. Reduce child mortality	Reduce by two-thirds between 1990 and 2015	Mortality rate for under fives and infants
5. Improve maternal health	Achieve attendance at 90% of deliveries by trained health personnel by 2015	% births attended by skilled personnel
6. Combat HIV/ AIDS, malaria and other diseases	50% reduction proportion of people without access to water between 1990 and 2015	% of population without access to an improved water source, death rates from diseases
7. Ensure environmental sustainability	Integrate environmental sustainability into national policies & programmes and reverse the loss of resources	% forest cover, energy use per unit GDP, biological diversity
8. Develop a global partnership for development	Free trade, good governance, debt relief, ODA	ODA that is not tied; debt reduction; telephone, Internet and computers per capita

Investments

Whether from or overseas, investment can be top down, e.g. Pergau Dam, trickle-down or bottom-up.

International initiatives

→ Earth Summit (1992 Rio and 2002 Johannesburg) and Agenda 21.
→ The Commission for Africa.
→ Drop the Debt.
→ Live8 and other grassroots initiatives.

Impact of Trade

Trade blocs can inhibit development or tie countries into poor trading conditions.

Fair Trade may provide higher incomes.

HIPC Initiative

This initiative led to the relief of debt according to a formula. 22 countries are receiving assistance (e.g. Malawi, Burkina Faso). Others such as Sudan have not met the criteria for debt relief.

Foreign Direct Investment

1. Aid dependency is measured as either aid as a porportion of GNI or aid per capita. **NGOs** have played an increasing role since the 1980s.
2. Role of World Bank, World Trade Organisation (WTO), International Monetary Fund (IMF).
3. Role of women in child health, the economy and as decision makers.
4. Self-sufficient balanced growth – China, India, Tanzania since 1960s is still being used.

Debt cancellation

The 1990s saw debt cancelled over a period of time on the condition that saved funds were spent on education and health. Strings are attached. MEDC taxpayers foot the bill.

Self-sufficient balanced growth

(e.g. China, India, Tanzania). This model was very popular among socialist states in the 1950s and 1960s, and is still being used in a modified form.

The jargon

Trickle-down is an approach to poverty alleviation: wealth will gradually filter down to the poor.
Bottom-up uses local initiatives: projects should be aimed at the poor, especially in rural areas and on small-scale projects.

Action point

Do the WTO and IMF work for the haves or have nots?

Action point

Is the Eurocentric view of women and gender equality the only one?

Checkpoint

Do you know the six main aims of Agenda 21?

Check the net

www.worldbank.org
is a vast site with many useful data sources. They also publish an atlas which is an excellent source of information.

Exam practice   answer: page 147

To what extent has international trade been responsible for the development gap? (25 mins)

Answers
Economic change

Globalisation

Checkpoints

1 The obvious Fordist industry has been car assembly lines. Other assembly line industries are white-goods manufacturing: fridges, washing machines, dish washers, vacuum cleaners.

2 IT and Biotechnology are named here. At a smaller scale, industries which offer bespoke goods such as designer fashion and indeed firms such as Benetton and Primark in the fashion trade shift production centres and types of goods on sale according to demand and the seasons

3 These could all apply to other RICs but the data is for India. *Benefits to India* are its 1.29 m graduates a year and strong IT in education system which provides young workers with higher incomes and higher starting salaries, up to 50% cheaper (wage of IT manager in USA $55,000 whereas $8,500 in India), and there is a multiplier effect, especially in Bangalore.

Costs to India are westernisation and loss of cultural identity, unsocial hours due to time differences, many employees leave after 3 years but without skills that can be transferred to other jobs in India, and it increases social divisions between rich and poor.

Benefits to the UK are improved profits of UK-based companies which benefit shareholders such as pension funds.

Costs to the UK are job losses and resentment among those who lose jobs leading to potential racism.

Exam practice

1 This is another question asking you for the higher order skill of evaluation. Therefore you should attempt to state the reasons from the text on pp132–3 but support them with examples to back up your case as the list has done. What the question demands and what is not on that list is some kind of evaluation of the relative weight of the factors that you have selected. The weight may vary from one industry to another. For instance car makers searched for global presence for the assembly of their brands. On the other hand chip makers searched for the cheapest labour cost sites for assembly. Your evaluation will score highly if you can show how the factors are all generally important but for some activities, some factors are more important. That is the route to A*.

2 The causes are; resources run out e.g. Cornish tin, are better elsewhere and are cheaper elsewhere; newer products have taken over; cheaper labour elsewhere; rationalisation; wring image of area for entrepreneurs – labour problems; neglected area as opposed to better sites elsewhere; taxation levels; global companies can switch production between countries.

You must state the region that you are looking at. South Wales or North East UK, the Ruhr are all good examples of declining industry, dereliction, government investment and new activities replacing the old.

Transnational corporations

Checkpoint

Most global companies have their HQs in the USA. 34% are US-based, but this is declining – 22% in Japan, 2% in Canada, and 7% in the UK, and declining. Most notable on the table is the rise of Chinese TNCs with 7% of the global total in 2008. Russia, India, South Korea and Mexico (not on table) have more Top 500 companies since 2003. The big companies are not those that geographers normally study as examples of TNCs: three companies are oil and gas producers, five are biotech companies and three are IT companies. The USA has the largest company in every economic sector except telecommunications and automobiles. Many TNCs are a force in Cultural Globalisation, e.g. Coca-Cola, Disney, Sony and Time Warner.

Exam practice

You could plan this answer as a table with a column of costs and a column of benefits. Much depends on the countries that you select because there is a further element of costs and benefits to the home, developed economies and the countries where major investments are being made, e.g. LEDCs, NICs and RICs.

An example of costs and benefits emanating from one company is Cargill – an agricultural trading TNC based in the USA. It supplies loans to farmers in Brazil to buy orange trees, purchases and processes the fruit, exports juice to the USA in its own supertankers, and sells to retailers. This has eliminated competitors in both Brazil and the USA.

Trade in food results in low prices for producers but profits for processors and retailers. For example, cocoa prices are much lower than the price of chocolate when it is re-exported to the cocoa-producing country. There will be environmental costs in agriculture of e.g. use of fertilisers and pesticides, the impact of monoculture, loss of biodiversity. However if the products are Fair Trade then the farmers will earn more money. There are social costs of westernisation because of low-grade service sector employment that serves western interests. The social benefit could be the increase in educational opportunities.

For MEDCs, look at the impact of deindustrialisation and the move of service sector work overseas.

Levels of development

Checkpoint

Crudely drawn line on the world map. NICs are put together with LEDCs – generalises across every LEDC from Mali to Singapore. It groups Ukraine with the USA and Germany. The reality is a gradation or a more complex set of types – former Communist Bloc, NICs, OPEC oil-rich countries. No account taken of size, e.g. small rich Kuwait as an LEDC. Line might be different if social development taken.

Exam practice

Discuss value of GNP, GDP, HDI, PQLI, labour force composition, exports by value, energy consumption, infant mortality, calorie intake, population per doctor, literacy, access to safe water as indicators with supporting evidence.

Look at regional GDP, social indicators – telephones per 000, unemployment, population per doctor, education levels.

Inequalities

Checkpoint

Six are African and 3 are Latin American and Haiti is in the Caribbean. Note that this table is a ranking of the worst and indicates that the highest are in the LEDCs but not always the worst if HDI is used as a measure. Most former Soviet Bloc countries have lower ratios and yet they do not have a high HDI.

Exam practice

The question does not ask you to cover all of the theories so you are able to select those that you know best. If you have memorised some diagrams this may help you both as an aid to understanding and as a means of gaining marks. For each model that you select, try to outline the good and bad points. It will impress the examiner if you can conclude by stating which model you favour and why. Remember that all models are only an approximation of reality and that they are all a product of the time when they were devised. Rostow is a simple explanation based on the temporal development of Europe and not necessarily applicable in countries which missed out stages such as many NICs.

Development gap

Checkpoint

Agenda 21 was agreed at the Rio Earth Summit 1992. There were six proposals for environmentally sound, sustainable development:

1 Support LEDCs in efforts to stop or modify projects known to harm the environment.
2 Give more resources to family planning, expand job opportunities for women so that population growth is slowed.
3 Allocate development assistance to programmes targeting the alleviation of poverty, environmental health and meeting basic needs for food and shelter.
4 Invest in research on energy alternatives to reduce greenhouse gases and slow climatic change.
5 Invest in research and agricultural advice to reduce soil erosion and develop more environmentally sensitive agricultural policies.
6 Provide funds to encourage biodiversity and protect natural habitats.

Exam practice

This is an evaluative question where you are expected to show how it has and how it has not. Trade does provide countries with income to develop but many TNCs are (or have been) driving down prices so that many get less for products such as bananas than they did a decade ago. Trade Blocs also restrict the ability of countries to penetrate markets and they also sign agreements with other trade blocs that do not permit free trade with outsiders. Trade has narrowed the gap through the migration of industrial jobs and the creation of new jobs in NICs such as Malaysia and RICs such as India and China. Offshoring of e.g. call centre work has helped some areas. However, the MEDCs now concentrate more on economic activities with higher order skills such as research and development, education and advanced manufacturing. Only when a product is developed might manufacturing be transferred to NICs.

147

Revision checklist
Economic change

1	Outline the impact of globalisation on contrasting areas of the world.	Confident	Not confident **Revise** page 136
2	Explain why globalisation has occurred.	Confident	Not confident **Revise** pages 136–7
3	Describe the growth of either a manufacturing or a service sector TNC.	Confident	Not confident **Revise** pages 138–9
4	Evaluate the strengths and weaknesses of the various methods of assessing levels of development.	Confident	Not confident **Revise** pages 140–1
5	Outline how trade and aid can assist in narrowing the development gap.	Confident	Not confident **Revise** pages 144–5
6	Evaluate the benefits to be gained from trickle-down and bottom-up approaches to development.	Confident	Not confident **Revise** page 145

Sustainability

Sustainable development is development which 'meets the needs of the present without compromising the ability of future generations to meet their own needs'. It is a process by which:

→ human potential (standards of well being) is improved and there is involvement in decision making
→ the environment (the resource base) is managed to supply humanity on a long-term basis
→ social justice is implied.

The definition suggests that mankind has degraded the planet and must make amends for future generations.

Exam themes

- Sustainable development
- Sustainable food supply
- Sustainable water supply
- Sustainable urban environments
- Sustainable environments

Topic checklist

	EDEXCEL		AQA		OCR		WJEC	
	AS	A2	AS	A2	AS	A2	AS	A2
Principles of sustainability	○		○		○			●
Sustainable environments					○			●
Sustainable food supply			○					●
Sustainable cities					○			●
Sustainable water supplies								●
Energy issues			○			●		●
Sustainable energy supplies			○			●		●
Sustainable tourism								●
Health and welfare			○					●
Waste disposal	○							●

Principles of sustainability

Strategic thinking – the WCED

The World Commission on Environment and Development (1987), chaired by Brundtland, investigated the capacity of the earth to support its population and the ways in which human activities were affecting the environment. These activities have created crises of environment, development, security and energy. Key aspects of the commission were:

→ the achievement of basic needs for all people, especially the global underclass
→ limits to growth are technical, cultural, and social.

Brundtland suggested seven major proposals for a strategy for sustainable development:

1. revive economic growth
2. change the quality of growth
3. meet basic needs of food, water, employment, energy and sanitation
4. stabilize population growth
5. conserve and enhance resources
6. adapt technology to manage risk
7. put environment into economics.

The Brundtland Commission reported to the United Nations General Assembly, which in turn requested a five-year progress report. This became the Rio conference, the largest environmental conference ever.

Managing global ecosystems – United Nations Conference on Environment and Development, 1992 (UNCED)

The outcomes of Rio were:

→ convention on biodiversity
→ framework convention on climatic change
→ principles of forest management
→ Agenda 21
→ the Rio Declaration on Environment and Development.

Local developments – role of local Agenda 21

As a result of the Earth Summit, national governments are obliged to formulate national plans or strategies for sustainable development. It is **people** who do development, not governments, and therefore sustainable development is a local activity. All people, however poor, have some ability (however constrained) of changing what they do, in small ways.

Local authorities are beginning to translate the global sustainability agenda into local action. Just as global sustainability cannot exist without national sustainable policies, national Agenda 21 is incomplete without local *Agenda 21*.

Check the net

One of the best sites is http://www.ace.mmu.ac.uk/esd/Principles/principles.html

Checkpoint 1

Define the term sustainable development.

Checkpoint 2

What is meant by the term Agenda 21?

Local authorities have a number of roles in sustainable development:

→ as a consumer of resources
→ as a force for change in the market place
→ as a role model for other organisations
→ as providers of information
→ as providers of services
→ as planners
→ as local governments and decision makers.

Concepts

Green growth refers to developments that are environmentally friendly and sustainable e.g. renewable forms of energy, low-technology agriculture and sustainable water management. It implies the use of the environment to improve standards of living for local people. Ecotourism in Nepal (see page 164) is a good example.

Economic sustainability refers to the development of economic activities that will last a long time since they do not destroy the world's resource base, and compromise the needs of future generations.

Environmental sustainability refers to the long-term sustainable use of environments and resources at present and in the future. A good example is how tropical rainforests or coral reefs can be managed to provide an economic return without destroying the resource base.

Action point

Find out what your local council is doing for its Agenda 21 commitments.

Checkpoint 3

What is meant by the term green growth?

Grade booster

Use specific examples – some from your locality – to show how sustainable development can be achieved.

". . . a process of change in which the exploitation of resources, the direction of investments, the orientation of technological development and institutional change are all in harmony and enhance both current and future potential to meet human needs and aspirations. It is possible to obtain sustainable development."

The World Commission on Environment and Development

Exam practice   answer: page 170

Consider the World Commission on Environment and Development's definition of sustainable development on the right. Assess how far the societies of the world have moved towards this target. Explain why there has or has not been progress. (15 mins)

Sustainable environments

Sustainable development in Curitiba

Curitiba, a city in south west Brazil, is an excellent model for sustainable urban development. It has experienced rapid population growth; from 300 000 in 1950 to over 2.1 million in 1990, but has managed to avoid all the problems normally associated with it. This success is largely due to innovative planning:

→ public transport is preferred over private cars
→ the environment is used rather than changed
→ cheap, low technology solutions are used rather than high technology ones
→ development occurs through the participation of citizens (bottom-up development) rather than top-down development (centralised planning).

Sustainable solutions to flooding

Problems (1950s/60s)

→ many streams had been covered to form underground canals which restricted water flow
→ houses and other buildings had been built too close to rivers
→ new buildings were built on poorly drained land on the periphery of the city
→ increase in roads and concrete surfaces accelerated run off.

Solutions (late 1960s onwards)

→ natural drainage was preserved – these natural flood plains are used as parks
→ certain low-lying areas are off-limits
→ parks have been extensively planted with trees, existing buildings have been converted into new sports and leisure facilities
→ bus routes and bicycle paths integrate the parks into the urban life of the city.

Transport

The approach to transport in Curitiba is very different. The road network and public transport system have structural axes. These allow the city to expand but keep shops, workplaces and homes closely linked.

There are five main axes of the three parallel roadways. The central road has two express bus lanes.

Curitiba's mass transport system is based on the bus. Inter-district and feeder bus routes complement the express bus lanes along the structural axes. Everything is geared towards the speed of journey and convenience of passengers.

→ a single fare allows transfer from express routes to interdistrict and local buses
→ extra wide doors allow passengers to crowd on quickly
→ double and triple length buses allow for rush hour loads.

Check the net

One of the best sites on sustainable environments is: http://www.brookes.ac.uk/schools/be/oisd/

Checkpoint 1

Where is Curitiba?

Checkpoint 2

How is flooding managed in Curitiba?

Checkpoint 3

What is the basis of public transport in Curitiba?

The rationale for the bus system was economic as well as driven by sustainability. A subway would have cost $70–$80 million per km, whereas the express bus ways were only $200,000 per km. The bus companies are paid by the kilometres of road they serve, not the number of passengers. This ensures that all areas of the city are served.

Sustainable urban development in LEDCs

Sustainable futures require:

→ use of appropriate technology, materials and design
→ acceptable minimum standards of living
→ social acceptability of projects
→ widespread public participation.

The main dimensions of sustainable development are:

→ provision of adequate shelter for all
→ improvement of human settlement management
→ sustainable land use planning and management
→ integrated provision of environmental infrastructure: water, sanitation, drainage and solid waste management
→ sustainable energy and transport systems
→ settlement planning in disaster prone areas
→ sustainable construction industry activities
→ meet the urban health challenge.

Sustainable development in rural areas

Sustainable development is defined as development that meets the needs of people at present (and improves basic standards of living) without compromising the needs of future generations. Sustainable development implies social justice and equity:

→ resource conservation
→ empowerment of local communities.

For most rural areas, sustainable development implies sustainable agriculture and/or sustainable tourism.

Rural aid for BARCA Province, Eritrea
→ Financial advice promoted by NGOs.
→ Establishment of three nurseries for demonstration projects.
→ 'Barefoot' doctors and provision of 'health for all'.
→ Widespread local participation and consultation.
→ Rural workshops to foster local industrialisation.
→ Integrated planning of farming, irrigation, soil and water conservation, reforestation, health and education.

Action point

In what ways is your local environment being made more sustainable?

Grade booster

Show that you are aware that attempts to achieve sustainability are not always successful. Find out examples, even from your local area, where attempts at sustainability have not succeeded.

Exam practice   answer: page 170

Evaluate the attempts to achieve sustainability for an environment/area that you have studied. (15 mins)

Sustainable food supply

Options for sustainable agriculture in LEDCs

Rural areas are especially important in LEDCs. They account for a large proportion of the workforce as well as providing food and export earnings. For about one-fifth of the population of LEDCs, environmental concerns and development needs are focused upon immediate survival.

A profile of agriculture in the Eastern Cape, South Africa

In general the Eastern Cape is an area of arable agriculture rather than pastoral farming. Most cultivation is of a subsistence nature. Agricultural productivity is low and variable for a number of reasons:

→ low and irregular rainfall
→ poor soils
→ limited vegetation quality
→ steep slopes
→ limited access to the land
→ high cost of seeds, water
→ lack of fences to deter domestic animals
→ shortage of labour
→ theft and the high risk of operations further reduce the incentive to farm the land.

Increasing agricultural productivity

There are many possibilities for improving agricultural productivity in the region. Irrigation schemes can be expensive but there are cheaper schemes, for example:

→ establishing small-scale gardens and subsistence farms – these could provide a very useful extra supply of food in just a few weeks. If such gardens are integrated with improvements in harvesting and storage, there ought to be a much larger supply of quality food at very little cost.
→ developing farming cooperatives – there is a pool of labour and experience, the latter perhaps somewhat limited, which could share the cost of tools and seed, and use the produce for their subsistence needs.
→ erection of fences or barbed wire – this would help prevent theft and trampling of crops by livestock. It could become a cottage industry.
→ using drought-resistant fodder crops – such as American Aloe. Pastureland is especially fragile owing to a combination of drought, overgrazing, population pressure, and the absence of land ownership policies. Trying to decrease herd size has proved unpopular and unsuccessful. Drought-resistant fodder crops are a good alternative.

Essential oils

The production of essential oils holds considerable potential as a form of sustainable agricultural development. Not only are the raw materials present but it is a labour intensive industry and would utilize a large supply of unemployed and underemployed people.

The essential oils industry has a number of advantages:

→ it is a new or additional source of income for many people

Check the net

One of the best sites is
http://www.sustainabletable.org

Checkpoint 1

Why are rural areas important in LEDCs?

Checkpoint 2

Why is agricultural productivity low in the Eastern Cape region of South Africa?

Checkpoint 3

In what ways are essential oils useful for agricultural development?

→ it is labour intensive and local in nature

→ many plants are already known and used by the people as medicines, and are therefore culturally acceptable

→ in their natural state the plants are not very palatable nor of great value and will not therefore be stolen

→ many species are looked on as weeds. Removing these regularly improves grazing potential as well as supplying raw materials for the essential oils industry.

Managing rural environmental problems generated by agriculture

Problems created by modern intensive farming include: the impacts of agro-chemicals such as eutrophication; soil erosion; desertification; hedgerow removal; loss of biodiversity; BSE; FMD (Foot and Mouth Disease); plus social problems such as those caused by the Green Revolution, e.g. the widening rich-poor gap in many LEDCs.

Action point

How is sustainable farming being achieved in your country?

Feature	Likely impact on landscape
Plant and animal breeding to produce higher yielding varieties	Has meant some agricultural land has become redundant and available for other uses, e.g. afforestation of hill land
Use of large quantities of fertiliser, especially nitrates, to increase yields	Problems of increased pollution of watercourses e.g. eutrophication
Use of pesticides (herbicides and insecticides) to cut down losses of crops	Loss of biodiversity of herb meadows Poisoning and killing of wildlife Many improved pastures lack diversity
Improvements in farm technology (e.g. more effective driers and harvesters)	Removal of hedgerows to make way for mechanisation Increased arable farming has led to massive soil erosion and 'cover' is removed
Changes in farming practice	Creation of large unsightly buildings for factory farming New extensification can lead to many hectares of hill land and the less-favoured areas being taken out of use

Grade booster

Bring in local food production – farmers' markets, production from allotments, local production – these are often examples of food production which is far more sustainable than globalised agribusiness.

Exam practice   answer: page 171

In what ways is globalised agribusiness bad for the environment? How can sustainable food production help? (15 mins)

Sustainable cities

Check the net

www.sustainablecities.net, www.monocle.com (has an annual survey of urban liveability), http://www.globalenergybasel.com, http://www.un.org/esa/sustdev/documents/agenda21/english/agenda21chapter28.htm, http://commitments.clintonglobalinitiative.org

Dimensions of sustainable cities

→ Economic sustainability – economy is able to sustain itself without affecting environment and resource base.

→ Social sustainability – policies to improve the quality of life and access to the natural and built environment by reducing poverty and increasing overall satisfaction.

→ Natural sustainability – management of resources and waste products, which depends on access to and rights over natural resources.

→ Physical sustainability – the ability of urban areas to support their population and its production.

→ Political sustainability – democratisation and participation of society in urban governance.

The Sustainable Cities Programme (SCP)

This is a joint UN-HABITAT/UNEP programme, started in the early 1990s, that operates in over 30 countries to package urban Environmental Planning and Management (EPM) approaches, technologies and know-how. It involves stakeholders in development strategies, mobilises local resources and commitment, and helps implement a) the Agenda 21 at the city level, b) the environmental component of the Habitat Agenda, c) the UN Declaration on Cities and other Human Settlements and d) the Millennium Declaration.

UK Sustainable Communities 2005

Sustainable communities should a) be environmentally sensitive, b) have good communications and transport linking people to jobs, schools and health, c) have a diverse, flourishing local economy, d) be well designed and built, e) be safe, tolerant and cohesive with strong local culture and community f) be well served by various services appropriate to needs. Thames Gateway is a case of sustainable urban development, involving several government agencies. The London Sustainable Development Commission, founded in 2002, has 20 quality-of-life indicators of sustainable development.

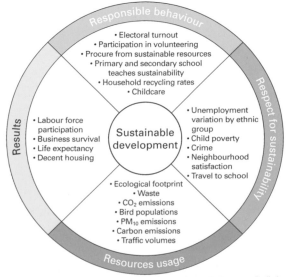

The Dimensions of city sustainability: Quality of life indicators

Source: amended from 'Sustainable Urban Development in the UK: Rhetoric or Reality', M.Pacione, published by Geographical Association, Autumn 2007

Ecotowns

Announced in 2008, these will be towns of up to 15,000 low carbon homes. The proposed sites for the towns to be completed by 2020 are Pennbury, Leics; Manby and Strubby, Lincs; Curborough, Staffs; Middle Quinton, Warks; Bordon, Hants; Weston Otmoor, Oxon; Ford, W. Sussex; Imerys China Clay Community, Cornwall; Rossington, S. Yorkshire; Cottishall, Norfolk; Hanley Grange, Cambs; Marston Vale and New Marston, Beds; Elsenham, Essex; and Rushcliffe, Notts. They are partly based on the idea initiated by HRH Prince Charles at Poundbury in Dorset.

City projects

Hammarby Sjöstad: The Sustainable City is an eco-friendly housing project in Stockholm that has been able to meet ambitious pollution and energy targets using a no-waste approach. It is a former "brown site", which has been redeveloped into an eco-friendly residential area. It has been obliged to recycle all of its waste to produce half the environmental impact of similar districts. The developers met the targets by integrating heat and electricity supply, waste management, waste water management, traffic, infra-structure, and urban planning. It will be completed by 2016 and have 10,000 apartments. Waste water is used to "harvest" bio-gas and manure, while the roughly purified water is used in the district's thermal power station which in turn, burns human solids bio-fuel.

There are also environmental efforts such as Urban Food supplies, Urban Gardens and roadside tree planting. The Clinton Global Initiative started a programme in 2007 in Bangladesh to try to make their low-lying cities more sustainable and proofed against flooding.

City projects also include transport efforts such as cycle ways (Bristol), congestion zones (Singapore and London) and public transport (Curatibo and Almere).

Checkpoint

Do you have examples of a large-scale brownfield development, plus a smaller-scale development, where you are able to measure the success of the efforts to be sustainable?

Action point

If you have an eco-town near you, keep a file of press cuttings which will provide you with information on why people need or are opposed to the new settlement

Grade booster

Try to develop your own definition of sustainability, rather than use Bruntland's. The dimensions listed here enable you to provide a more complex, higher level definition.

Sector	Sustainability Indicators
Transport	car use, number of short journeys, freight traffic
Leisure & Tourism	leisure journeys
Energy	energy consumption, use of fossil fuels, renewable fuel use
Land Use	use of derelict land brown field sites, urban development, green spaces
Water Resources	rainfall, demand & supply of public water
Climate Change	greenhouse gas emissions
Ozone Layer Depletion	CFCs consumption
Acid Rain	road transport emissions of sulphur dioxide and oxides of nitrogen
Air Quality	pollutant emissions, money spent on air pollution reduction
Freshwater Quality	chemical and biological river quality, pollution incidents, money spent on sewage treatment
Coasts	bathing water quality, oil spills and discharges
Minerals Extraction	mineral workings on land, amounts of waste
Waste	household & industrial waste, recycling, landfill waste

Exam practice answers: page 171

1 Why is it difficult to make existing cities sustainable? (10 mins)

2 Outline and evaluate a scheme that attempts to deliver a part of Agenda 21 in your local area. (20 mins)

Sustainable water supplies

It has been estimated that as many as two fifths of the world's population are short of water. Many of the reservoirs are regularly at some of the lowest levels since records began. Floods and droughts are happening more frequently with public drought orders being a common event in parts of the UK; it seems very likely that global warming is to blame for these effects on our water supplies. As water shortages increase there are likely to be many consequences.

Checkpoint 1

What does the term sustainable mean?

Consequences of increasingly limited water supplies

As water supplies become more limited the following consequences could occur more frequently:

→ Migrations as people move to find reliable water supplies.
→ Food shortages as peoples' crops fail due to water shortages.
→ Starvation as a consequence of severe water shortages and crop failure.
→ War as people fight over a declining resource.

Causes of increasingly limited water supplies

→ Population growth. As the world's population continues to grow, so does the demand for water.
→ Economic development. As countries and regions improve their level of development, people are using more water both in their homes and at work and the supply of sufficient quantities are becoming a challenge for the governments. This is particularly noticeable in the rapidly growing economies of China and India.
→ The mismanagement of supplies allowing wastage or pollution of supplies to take place.

Checkpoint 2

Explain how economic development leads to a greater demand for water supplies.

Managing water supplies in a sustainable manner

There are a number of ways that water supply and demand can be managed sustainably, these include:

Action point

Research the Three Gorges dam in China as an example of a HEP scheme that has been designed to allow greater sustainability of water supply in China.

→ Building new storage facilities. These always have economic, social and environmental issues and are only possible after lengthy planning proposals.
→ The development of new groundwater sources by digging new wells.
→ Water transfers from areas of surplus to areas of deficit. In the UK this is a transfer from the North and West to the South and East.
→ 'Grey' water. This is water that has already been used for such things as bathing, washing clothes etc and can be used for another purpose such as irrigation and toilet flushing. This has the double advantage of reducing the amount of water that is being sent for sewage or septic tank treatment, and of reducing the demand for fresh water.
→ Recycling water, for example in agriculture where modern glasshouses reuse the same water several times.
→ Desalination. An energy-expensive alternative by which fresh water is obtained from seawater.
→ Reducing consumption in a number of ways, some examples of

which follow.

→ Use water more efficiently at home and work by:
  → Turning water off whilst brushing teeth, which has been estimated to save as much as 5 litres a minute
  → Using hosepipes as little as possible
  → Using washing machines efficiently by having full loads
  → Recycling water in the kitchen where possible
  → Having a shower rather than a bath
  → Repairing drips as soon as possible
  → Reducing the volume of water in the toilet cistern
  → Washing up in a bowl, not the sink
  → Not using dishwashers as they waste water and energy.
→ Encouraging institutions and companies to save water by adopting many of the same strategies.
→ Pressuring governments to make better use of water and introduce legislation to protect future supplies.
→ Getting more efficient use from water supply companies as lots of fresh water is wasted through leaks in supply routes and leaking reservoirs. In some countries laws and directives exist to encourage a more efficient supply industry.

Problems of managing water supplies in a sustainable manner

Supply is exceeding demand in many parts of the world.
Water resources are often moved across political boundaries. The negotiations about supplies and the vulnerability of a nation relying on another for its water can be difficult political issues.
In some areas people treat water as a disposable commodity and are very wasteful. A change of attitude is needed to reverse this trend.
The uncertainty of water supplies resulting from climatic and other changes being brought about by global warming are making it more and more difficult to manage supplies.

Checkpoint 3

Desalination of seawater has been proposed as a method of supplying fresh water in some parts of the world. Explain how this works and why it is not a realistic proposition in many countries.

Action point

The waters of the river Jordan are used to supply water to a number of countries. Research the supply systems and explain how this might cause international conflict.

Grade booster

Examiners will be hoping to read answers which clearly demonstrate how managing water supplies is becoming a growing issue as other changes are taking place on the planet. Ensure you are able to make links clearly with economic change as well as global warming and population growth.

Exam practice   answer: page 171

Explain how water supplies might be managed in a sustainable way in the future. (20 mins)

Energy issues

Energy is essential to human life. It provides heat, light and power and is the life blood of society. Energy resources are getting scarcer and people are having to conserve energy and discover new sources so that future generations will have a sustainable future. There is a politics of energy resulting from our efforts to obtain energy which may involve local conflict or, eventually, global conflict.

Energy types

Checkpoint

Are the following primary or secondary energy sources: wood, diesel, solar, wave?

Primary energy includes all basic sources, e.g. coal, oil, solar, wind. Secondary energy is produced from the primary source, e.g. electricity, petrol.

There are four main types of energy: **mechanical** – clockwork; **electromagnetic** – electrical energy; **chemical** – photosynthesis; **nuclear** – radioactive fission. Energy flows through an economic system as a natural resource. It is a raw material for industry, agriculture and transport, and provides heat and light.

Fossil fuels

Examiner's secrets

Only attempt projects on energy if you are able to look at the environmental effects of, for example, a wind farm site or proposal.

Fossil fuels form about 75% of global energy use (more in MEDCs).

Oil 35% of primary energy consumption in UK. Use falling in MEDCs and rising in LEDCs. 54% used in transport, 20% in homes and 19% in industry. Production dominated by the Middle East, so the politics of production should be understood.

Coal 16% of primary energy consumption in UK. Use in UK has been affected by politics, e.g. miners' strike. Large reserves – UK reserves could last 1000 years. Use is declining due to pollution and CO_2 issues. World reserves mainly concentrated in USA, Russia and China.

Natural gas 38% of UK primary energy. Main sources are Norway and Russia. World production dominated by North America, Europe, Russia. Reserves are twice those of oil and found in many LEDCs and NICs. Some producers are now searching for substitutes, e.g. shale and coal-seam methane in the USA.

Nuclear power 9% of UK primary energy. 90% generated in MEDCs. Future use controversial because high cost of building/decommissioning, disposal of spent fuel, impact of disasters Chernobyl. In 2007 UK government reversed previous policy and is now supporting the expansion of nuclear generating capacity.

Hydro-electricity 2% of primary energy in UK. Mainly produced in MEDCs although some notable dam schemes provide power for LEDCs, e.g. Kariba, Akosombo and Aswan dams.

Biomass

Grade booster

Answers of the highest quality will show how examples of energy schemes, such as dams, will have an impact on people, the economy and the environment.

Biomass accounts for 6% of global energy consumption, of which wood is the main source. Some 21 countries depend on wood for 75% of their energy (it is 90% in Mali and Burkino Faso). Animal dung can be up to 90% of energy consumption in Indian villages. Straw is also a source.

Fuelwood gathering takes 10% of Peruvian women's time; in Kenya women spend 24 hours a week collecting firewood.

The politics of energy

As the world's population is increasing and economies of the world are developing, there is a growing demand for energy of all types. Fuels are needed to power the growing number of factories that are providing the goods for the increasing number of consumers and to heat and light their homes. Fuels are also required for the vehicles used to move the increasing number of people and goods around the world. The recent rapid economic growth in China and India is leading to a sudden increase in demand for many resources, particularly energy sources, with the result that the price of oil has risen considerably and there has been an increase in the use of alternative sources. The increasing use of biofuels produced from crops grown on land which had previously been used to produce food, has been suggested as a factor in the rising cost of food in many parts of the world.

Oil remains a very valuable global commodity and its supply continues to be a concern to those countries that depend upon oil imports to power their country. Some of the key issues regarding global oil production are:

→ The growing concerns about the pollution caused by its continued use.
→ Concerns about the reliability of supplies from politically unstable countries, particularly in the Middle East.
→ The continuing shortages of energy supplies faced by many people in the Less Economically Developed countries of the world.
→ The influence that Russia has over the economies of Western Europe through their reliance on Russian oil.

Exam practice   answer: pages 171–172

1 What is the global energy crisis? (8 mins)

2 How is the uneven distribution of energy resources affecting the economic development and energy policy of one LEDC and one MEDC? (20 mins)

Sustainable energy supplies

Energy is used by all people whatever their level of development and is essential to human life. Our present energy consumption is NOT sustainable. Many of the energy resources we are using are finite, and present energy consumption is leading to environmental damage through pollution and is implicated in global warming. If sustainability of energy consumption is to be achieved, there will have to be greater conservation of supplies, research into alternative sources, and improvements made to pollution control.

Problems associated with the supply of energy

There are many reasons why the supply of energy is not sustainable today. **Environmental** problems include the increasing carbon footprints of people as their lifestyles and quality of life improve. The increasing CO_2 levels have played a substantial part in Global Warming with all its related problems. The generation of electricity in fossil fuel burning power stations has increased sulphur dioxide levels in the atmosphere and caused environmental damage through acid rain. HEP schemes have altered the ecosystems and drainage patterns in river basins.

Economic problems are things such as the enormous costs of disposing of nuclear waste and radioactive materials following the closure of a nuclear power station. The use of farmland in countries such as Brazil to produce bio fuels instead of food has contributed to the increasing cost of food. The rising cost of fuels as the supplies become more scarce. The debts that some LEDCs are saddled with as a result of building HEP schemes such as Itaipu on Parana River in Brazil.

Political issues include whether or not to allow the construction of a new power station; the debate about nuclear plants and their safety and waste disposal problems; the environmental damage that would be caused by the construction of a new HEP scheme; the degree of vulnerability a nation has to political decisions in other countries when they rely on imported energy supplies, and the local objections that can occur about the building of a new wind farm.

Technological problems can include the design of environmentally efficient vehicles and power stations; the development of pollution control systems; the silting up of major HEP schemes.

The changing demand for energy

Demands for energy are changing in quantity and type.

→ The growth in air transport has been dramatic in the last 30 years resulting in a growing demand for kerosene, which is refined from crude oil.

→ As people's quality of life improves, they tend to use more energy. For example they may heat their homes or cook with oil, when previously they had used locally collected firewood.

→ The rapid economic development in NICs has resulted in a growing demand for all sources of energy for industries, transport systems and domestic use.

→ As technological advances are made, new devices are invented which often use a source of energy rather than manual labour.

Managing the demand for energy sustainably

This means balancing socio-economic and environmental needs and requires detailed planning and management.

→ Continue to research into and develop alternative sources of energy.

→ Aim to use energy more efficiently wherever possible.

→ Change people's habits about energy consumption and in this way reduce demand.

→ Continue to promote energy efficiency on white goods and cars.

→ Promote more international cooperation about energy policies and carbon emissions.

→ Encourage reliable schemes of carbon trading.

→ Ensure that transport improvements are cost effective in both an economic and environmental sense.

→ Design and promote greater energy efficiency in home building.

→ Encourage and facilitate car-pooling.

→ Develop more video conferencing to reduce business travel.

→ Encourage home working using ICT rather than traveling to offices everyday.

→ Promote and develop more sources of renewable energy (see below).

Action point

Read the summary of the Eddington Report (December 2006) to understand the recommendations to government about future transport policy.

The jargon

White goods means household appliances such as fridges, microwave ovens and washing machines. They are now marked with energy efficiency information clearly displayed.

Checkpoint 2

What is carbon trading?

Renewable energy

→ Solar power needs high sunshine totals, e.g. deserts, Pyrenees.

→ Modern wind generation needs strong winds, e.g. Palm Springs, California.

→ Wave power needs a coastal region: still under development.

→ Tidal barrages need a large tidal range, e.g. Rance estuary, Brittany.

→ Geothermal only where there are igneous rocks, e.g. Larderello, Italy and New Zealand.

→ These all have an environmental impact.

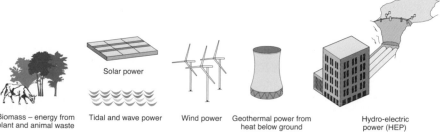

Biomass – energy from plant and animal waste   Tidal and wave power   Solar power   Wind power   Geothermal power from heat below ground   Hydro-electric power (HEP)

Examiner's secrets

Make sure you know a case study of the environmental effects of a dam, e.g. the Aswan dam, on river flow, irrigation, and the Nile delta. What are the benefits and drawbacks of the scheme? The Colorado is another good example.

Exam practice   answers: page 172

1   Can a sustainable energy supply be maintained in the future? (20 mins)

2   'The viability of renewable energy resources depends on physical geography and not the ability to develop them.' Discuss this statement. (15 mins)

Sustainable tourism

In 1993 the Northern Ireland Tourist Board (NITB) published the report *Tourism in Northern Ireland: A Sustainable Approach*. The main development priorities were seen as:
1. relating tourism to the environment
2. involving the local community
3. the need to aim for quality
4. the need for all parties working together.

Check the net

Two good sites include
http://www.tourismconcern.org.uk and
http://www.thetravelfoundation.org.uk/
cyprus.asp
(for an example from Cyprus)

Sustainable tourism: the benefits

Benefits for the visitor
→ The development of a quality tourism service.
→ Better relationships with the local community.
→ Closer involvement with and better understanding of both the people and the holiday destination.

Benefits to the tourism industry
→ Economic benefits for operators – e.g. reducing energy by installing efficient insulation.
→ Enhanced appeal of Northern Ireland for visitors from market areas that have ecologically-aware consumers.
→ Opportunities for the development and promotion of environment-friendly activity tourism e.g. conservation holidays.

Benefits to the environment
→ Safeguarding the resource for the benefit of future generations.
→ The protection and enhancement of the special landscapes which form much of Northern Ireland's appeal to visitors.

Benefits to the community
→ Opportunities for community involvement in tourism and the creation of a better climate for development.
→ Supporting the local economy and local services e.g. helping to support local transport systems in rural areas.
→ Creation of new business opportunities.

Ecotourism in Nepal

Deforestation on the steep slopes of the Himalayas in Nepal has led to severe erosion, removal of topsoil, and flooding. The Nepalese government has encouraged a number of community reforestation schemes, one of which is the Annapurna Conservation Area Project (ACAP). It has four distinctive characteristics:

1. it covers a large area, over 2200 km^2, and involves over 40,000 people
2. it is managed by local people since the government believe that it will only work if it involves and benefits the local community
3. it is based on landuse zoning
4. part of the funding comes from the tourist trade.

Mountain tourism

There are few roads in the Nepalese Himalayas so tourists must walk, or 'trek', to enjoy the mountains. Except on the more popular routes, trails are in poor condition; acute deforestation has led to frequent landslides; boarding and lodging facilities are poor; and clean drinking water is not always available. Yet, because tourism is the country's second highest foreign exchange earner (after carpet exports), contributing nearly 4% of gross domestic product, the government is keen to attract more tourists.

The Annapurna Conservation Area Project – a success story?

Village control of forests

ACAP was established in 1986. The conservation area does not exclude human habitation; the needs of the local people are integral to the project. After initial scepticism, local people have come to realise that they have a lot to gain from the situation.

The traditional method of managing Nepal's forests (known as ritithi) is balanced cutting and growth. However, the practice died out and decades of deforestation ensued as the population grew. Since the formation of ACAP, however, the responsibility of managing forests and wildlife has been handed back to the villagers.

New ideas

Fees paid by trekkers passing through the Annapurnas fund ACAP. It has devised a multitude of simple fuel-saving devices, including solar-heated showers, and runs health-related workshops. These workshops include hygiene (the necessity of boiling water or digging latrines) and using flues to prevent buildings filling up with smoke and contributing to one of the biggest health problems, tuberculosis.

Consequences?

The Annapurna Conservation Area is visited by more than 80,000 people a year. Over 700 tea shops and lodges have been built. Today, the local people are learning to maintain a high level of control over their resources and their future by building an endowment fund from entry fees to the area and working together on education, community development, biodiversity, and energy conservation projects.

The concern of mountain people is that they will lose control over their culture, their economy, and their environment as tourism develops. Merely restricting tourism cannot be the solution to this imbalance, because people's desire to see new places will not disappear. Instead, mountain communities must achieve greater control over the rate at which tourism grows.

> **Checkpoint 1**
>
> What are the benefits of sustainable tourism?

> **Checkpoint 2**
>
> What are the four distinctive characteristics of the Annapurna Conservation Area Project in Nepal?

> **Checkpoint 3**
>
> How have the fees paid by trekkers benefited local people?

> **Action point**
>
> How is sustainable tourism being achieved in your country?

> **Examiner's secrets**
>
> Named examples and places always impress examiners – but make sure you get your facts right.

Exam practice   answer: pages 172–173

Evaluate the success of the Annapurna Conservation Area Project. (15 mins)

Health and welfare

Health and welfare considers economic, social and environmental factors that influence a population's standard of living and quality of life. It has close links to population and economic geography and is important in the study of development and sustainability. The topic identifies differences in levels of health and welfare by comparing and contrasting conditions in different places. It considers a range of causal factors and how these interlink. An underlying theme is that, in their provision for health and welfare, countries need to adapt to constantly changing economic and social conditions.

Origin and transmission of diseases

The figure below compares the causes of death in Africa and Europe.

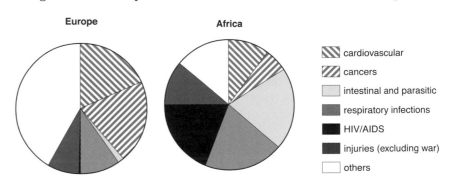

Types of transmission

→ Direct contact through blood or sexual fluids, e.g. HIV, syphilis, hepatitis.
→ Contaminated or insufficient water, e.g. cholera, typhoid, leprosy.
→ Insect vector, e.g. mosquito, tsetse fly.
→ Airborne in droplets, e.g. TB, whooping cough.

Cholera is common in poor and overcrowded conditions, and an **epidemic** (rapid spread affecting large numbers) often occurs after severe floods and cyclones, leading to diarrhoea, vomiting, thirst, and muscle cramps.

Impacts of disease

Disease impacts on society both socially and economically. Wealth plays a key role in a country's ability to prevent and treat disease.

Reducing the impact of disease – LEDCs

Disease reduces the workforce and consequently productivity. The costs of response are high and cause social stress. A combination of preventative and medical measures is necessary to tackle disease. LEDCs benefit from the transfer of technology in the form of medicines, vaccinations and equipment, enabling them to reduce the incidence of disease despite their own limited income.

Some monetary measures involve spending on:

→ Education and training – leads to better treatment and fewer die.
→ Improved infrastructure – better housing, sewage and clean water, so disease less likely to occur and spread. Fewer cases, impact of disease reduced.
→ Vaccinations – protection against childhood diseases. Fewer cases, impact of disease reduced.

Improved nutrition increases resistance to disease, and measures like the spraying of mosquito-infested areas can be effective. However, the control of disease requires ongoing high cost attention which countries cannot always meet. There are many entirely preventable diseases concentrated in LEDCs that inhibit economic development.

Potential impacts of disease – AIDS in LEDCs

As yet there is no cure for AIDS, although life-prolonging drugs are available. It is spread by direct sexual contact with an infected person and, to a lesser extent, by contaminated blood in transfusions or open wounds. 80% of all AIDS cases occur in Africa where sexual intercourse without the use of a barrier method of contraception is common. It can also be passed from mother to baby. Deaths from AIDS in 1998 totalled only 4.2% of mortalities. The potential for an epidemic in future is high because the long incubation period hides the extent of its spread. Over 20% of the populations of Zimbabwe and Zambia are thought to already be infected with HIV. If an epidemic occurs the workforce will be reduced and economic development held back. Less capital will be earned and so less money will be available for prevention and treatment. Thus the spread is likely to continue, further holding back the country's development.

Don't forget!

Poor water supplies cause over three-quarters of all diseases in LEDCs. (Source: World Health Organisation)

The jargon

Remember to distinguish the HIV virus (Human Immuno–Deficiency Virus) which can lead to Acquired Immune Deficiency Syndrome (AIDS).

Checkpoint 1

Using a flow diagram, show the potential effect of AIDS on a country.

Checkpoint 2

Can you outline the impacts of disease in MEDCs?

Examiner's secrets

Make sure you give examples to support your point. Try to identify regions of the world that have the most acute problems. It would be good to point out that rapid natural increase makes responding to the problems harder too. The examiner would also credit a point that tackling disease is not always a question of lack of wealth but more one of lifestyle and social behaviour.

Exam practice answers: page 173

1 With reference to specific examples, outline the ways in which diseases can be spread. (8 mins)

2 Explain how a lack of wealth can affect a country's ability to tackle disease. (15 mins)

3 Outline the main economic effects of disease. (10 mins)

Waste disposal

Check the net

Two good sites include:
http://www.ace.mmu.ac.uk/eae/Sustainability/Older/Waste_Disposal.html
http://www.audit-scotland.gov.uk/docs/central/2007/nr_070920_waste_management.pdf

The UK has a poor record of disposing of waste. Recycling refers to the processing of industrial and household waste (such as paper, glass, and some metals and plastics) so that materials can be reused. This saves scarce raw materials and helps reduce pollution. The UK, for example, has long lagged behind with recycling, mainly because there are many more landfill sites which are cheaper to use. The UK has a recycling target of 33% by 2015.

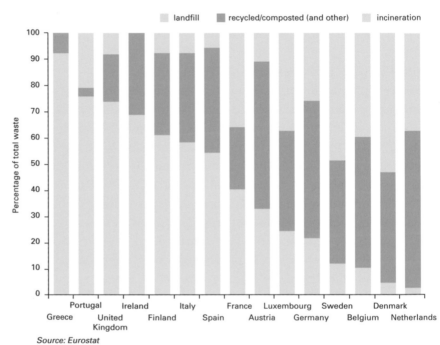

Municipal waste management in the European Union

Source: Eurostat

Checkpoint 1

What are the three Rs of waste disposal?

Checkpoint 2

Give an example of substitution.

Checkpoint 3

Why are landfill sites so common in the UK?

Action point

How is waste managed in your school or college?

Re-use refers to the multiple use of a product, by returning it to the manufacturer or processor each time. Re-use is usually more energy- and resource-efficient than recycling.

Reduction (or 'reduce') refers to using less energy, such as turning off lights when not needed, or using only the amount of water needed when boiling a kettle.

Substitution refers to using one resource rather than another – the use of renewable resources rather than non-renewable resources would be of major benefit to the environment.

Landfill is the burying of waste in the ground, and then covering over the filled pit with soil and other material. Landfill may be cheap but it is not always healthy – and will eventually run out. Most landfill is domestic waste but a small amount of hazardous waste is allowed on general sites.

Fly-tipping is when people or companies dump waste and old equipment – this is an increasing problem. There are many reasons for the increase:

➜ increased costs of landfill

→ more goods such as TVs, computers and refrigerators classified as 'hazardous' and subject to restrictions on how they are disposed of

→ the introduction of strict new EU regulations mean that a high proportion of new products must be recycled.

This can be costly to manufacturers and purchasers.

Waste imports in China

A fairly new environmental problem is the dumping of old computer equipment. To make a new PC requires at least 10 times its weight in fossil fuels and chemicals. This can be as high as 240 kg of fossil fuels, 22 kg of chemicals and 1500 kg of clean water. Old PCs are often shipped to LEDCs for the recycling of small quantities of copper, gold and silver. PCs are placed in baths of acid to strip metals from the circuit boards, a process highly damaging to the environment and to the workers who carry it out. China imports more than 3 million tonnes of waste plastic and 15 million tonnes of paper and cardboard each year. Containers arrive in the UK and other countries with goods exported from China, and load up with waste products for the journey back. A third of the UK's waste plastic and paper (200,000 tonnes of plastic rubbish and 500,000 tonnes of paper) is exported to China each year. Low wages and a large workforce mean that this waste can be sorted much more cheaply in China, despite the distance it has to be transported.

China is increasingly aware that this is not 'responsible recycling' and that countries are exporting their pollution to them. They have begun to impose stricter laws on what types of waste can be imported.

> **Grade booster**
>
> There are advantages and disadvantages to sustainable waste disposal. Be aware of the disadvantages – such as cost – as well as the advantages.

Exam practice   answer: pages 173–174

To what extent is the recycling of computers sustainable? (15 mins)

Answers
Sustainability

Principles of sustainability

Checkpoints

1 Sustainable development refers to the improvement in the quality of living without compromising the quality of life of future generations. The World Commission on Environment and Development (WCED) chaired by Brundtland, investigated the capacity of the earth to support its population and the ways in which human activities were affecting the environment.
Brundtland suggested seven major proposals for a strategy for sustainable development:
  1 revive economic growth
  2 change the quality of growth
  3 meet basic needs of food, water, employment, energy and sanitation
  4 stabilise population growth
  5 conserve and enhance resources
  6 adapt technology to manage risk
  7 put environment into economics.

2 As a result of the Earth Summit, national governments are obliged to formulate national plans or strategies for sustainable development – these are called Agenda 21 strategies. It is people who do development, not governments, and therefore sustainable development is a local activity. All people, however poor, have some ability, however constrained, of changing what they do, in small ways.

3 Green growth refers to developments that are environmentally friendly and sustainable, e.g. renewable forms of energy, low-technology agriculture and sustainable water management. It implies the use of the environment to improve standards of living for local people.

Exam practice

A question like this requires you to give some examples of ways that countries are moving towards sustainable development and obstacles to this progress and then sum up the overall state of affairs as you see it. The reasons for the progress and lack of it will probably be incorporated quite easily in what you write. You could include trade agreements, debt cancellation, population changes, 'green' policies, such as recycling, organic farming and renewable energy. You should also mention the increasing trend towards global cooperation as exemplified by the Rio and Tokyo summits as well as the less successful Hague summit of 2000.

Sustainable environments

Checkpoints

1 Curitiba is in south west Brazil.
2 Sustainable solutions to flooding include natural flood plains used as parks; parks have been extensively planted with trees to increase interception; and certain low-lying areas are not used for development.

3 Curitiba's mass transport system is based on the bus. Inter-district and feeder bus routes complement the express bus lanes along the structural axes.

Exam practice

The wording asks for an evaluation. Curitiba is considered to be one of the best examples of sustainability at an urban level. For example, in terms of transport, public transport is geared towards the speed of journey and convenience of passengers:

→ a single fare allows transfer from express routes to interdistrict and local buses;

→ extra wide doors allow passengers to crowd on quickly;

→ double and triple length buses allow for rush-hour loads.

The rationale for the bus system was economic as well as sustainability. A subway would have cost $70–$80 million per km - the express bus ways were only $200,000 per km. The bus companies are paid by the kilometres of road they serve not the number of passengers. This ensures that all areas of the city are served.

The system has been a success although it has not been replicated in much larger cities – either in Brazil or in other countries.

Sustainable food supply

Checkpoints

1 Rural areas are especially important in LEDCs. They account for a large proportion of the workforce as well as providing food and export earnings. For about one-fifth of the population of LEDCs, environmental concerns and development needs are focused upon immediate survival.

2 Agricultural productivity is low and variable for a number of reasons: low and irregular rainfall; poor soils; limited vegetation quality; steep slopes; limited access to the land; high cost of seeds and water; lack of fences to deter domestic animals; shortage of labour, and theft and the high risk of operations further reduce the incentive to farm the land.

3 The production of essential oils holds considerable potential as a form of sustainable agricultural development. Not only are the raw materials present but it is a labour intensive industry and would utilize a large supply of unemployed and underemployed people. The essential oils industry has a number of advantages: it is a new or additional source of income for many people; it is labour-intensive and local in nature; many plants are already known and used by the people as medicines, and are therefore culturally acceptable; in their natural state the plants are not very palatable nor of great value and will not therefore be stolen; many species are looked on as weeds. Removing these regularly improves grazing potential as well as supplying raw materials for the essential oils industry.

Exam practice

There are a number of ways in which globalised agribusiness bad for the environment. Examples include (and should be expanded upon) – the increased use of fertilisers (leading to eutrophication); food miles and release of fossil fuels leading to global warming; the spread of disease in monocultures; removal of vegetation (hedgerow removal and deforestation) leading to soil erosion and flooding. NB the question is about environmental impacts.

Sustainable food production can help by not removing vegetation; not using chemical fertilisers; supporting local markets so not using excessive air-miles, and having a varied agricultural basis so the spread of disease is reduced.

Sustainable cities

Checkpoint

Any of the Docklands redevelopments could be useful but remember that you have to say whether it has been successful. The small scale development could be for example, the attempts to clean up the River Wandle in South London so that it can be used for fishing and recreation and leisure rather than a dumping ground for rubbish. Success is whether the projects have managed to make the environment more sustainable. Some may not have been successful for a variety of reasons.

Exam practice

1 You could definitely develop the theme that cities were developed before sustainability came to the top of the agenda in the last two decades. Therefore, managers are dealing with developments over the centuries such as putting small streams underground, poorly insulated buildings and road schemes to name just a few. There has to be a will for organisations to cooperate and some might drag their feet. Some local governments can also be slow if they are of a political make up that resists sustainability. It costs money which has to come from somewhere such as taxes, sponsorship or personal involvement.

2 The table gives you a list of potential topics and you would need to select one that can be found near you. The question does ask you to state whether it is a success by evaluating it. Your evaluation might use the opinions of groups, reactions of companies and local government. For instance, in Bedhampton there are 'Bedhampton volunteers', who have clubbed together to maintain roadside verges and roundabouts by flower planting, rubbish clearing and placing and maintaining hanging baskets on lampposts. This meets with council approval because it saves on taxes because the council is not paying. It is a community grass roots initiative which creates local pride. On the other hand some feel that it is merely a form of political involvement because it is organised by Conservative councillors and that a local authority should be expected to maintain its own land.

Sustainable water supplies

Checkpoints

1 Sustainable means using a resource in a way that it will still be available in the future. A closely associated term is 'sustainable development', which means meeting the needs of the present without damaging the needs of the future.

2 Economic development leads to a greater demand for water supplies through improved public sanitation, new industries using water in their processes, the finance becoming available to pay for irrigation systems, more water-using devices being purchased for homes and greater expectations of a reliable public water supply network as economies develop.

3 Desalination is the heating of seawater, and its subsequent condensation to produce fresh water. This is a very energy consuming process as heat is needed to heat the water and energy is often needed to cool down the vapour as well. The high-energy costs have made this process uneconomic in most parts of the world. The United Arab Emirates lead the world in commercial desalination to produce fresh water.

Exam practice

This answer requires you to explain as many ideas as you can, about how water supplies can be sustainably managed. You should be able to refer to examples of new water storage facilities being built; methods of reducing consumption; the increasing use of "grey" water; more recycling of water, particularly in agriculture; desalination and its increasing potential as technologies improve; water transfers between drainage basins to even out supply and demand and the continued management of groundwater sources.

Energy issues

Checkpoint

Wood, solar and wave are primary, diesel is secondary.

Exam practice

1 It is the declining reserves of fossil fuels in the case of MEDCs and the decline of fuelwood in many LEDCs. It is a creeping crisis because as shortages arrive so prices rise and only the richest countries can afford energy. Crisis also caused by waste and inefficiency. Recently the issue of biofuels being grown on cropland has added to the crisis.

2 You can choose your countries.
LEDC. For countries without fossil fuels it might mean increased consumption of renewable timber but this can prevent regrowth, e.g. Nepal. High cost of other fuels can lead to debt. For those with some fossil fuels exported to pay debts this can lead to not investing in the future, e.g. Nigeria. The theme is one of slowing economic growth and even stagnation and 'negative growth'.
MEDC. Might have to invest in alternative supplies or expensive atomic energy (e.g. France, which has few home resources as coal and small oil and gas reserves run out).

Sustainable energy supplies

Checkpoints

1 Acid rain is any type of precipitation that has an unusually low pH. It is usually caused as a result of sulphur dioxide and nitrogen dioxide being emitted into the atmosphere by industry or vehicles and then being absorbed into clouds before being dropped onto the surface of the earth.

2 Carbon trading is a system whereby companies or individuals pay an organisation to compensate for the CO_2 they produce by operating schemes such as reforestation, waste reduction and renewable energy supplies. New low carbon technologies are being financed partly through carbon trading.

Exam practice

1 Here you should discuss reasons for and against the idea that sustainable energy can be maintained. In favour of the argument are ideas such as the growing research into alternative fuels and new technologies, international cooperation and treaties being signed, energy conservation and recycling policies which are reducing demand, new "green" taxes, more renewable sources being used and the idea of carbon trading. The pessimistic view contains arguments such as seemingly continual growth in demand, especially from NICs, dwindling of supplies of fossil fuels, the shortage of space for new HEP schemes, political insecurities in areas of fossil fuel supply (e.g. Middle East and Russia), a slow appreciation of the need for conservation strategies in MEDCs.

2 You should deal with two if not three sources. **Solar** depends on good sunshine records throughout the year. Cost is also a deterrent, so there is support for the alternative view. **Tidal** needs an estuary with a good tidal range. Very few attempted, e.g. Rance, because of costs. There is an environmental issue to do with altering the physical geography of estuaries. **Wave power** is costly, and has an environmental impact on the area. Probably viable but other factors rule it out at present. **Wind** is unsightly although there is a magic to wind farms, e.g. Palm Springs. Again, needs constant wind, so sites stand out. Costly to develop and still only supply relatively small amounts of power. **Geothermal** still needs power stations. Possibly the most constant, e.g. Larderello in Italy. All need to be backed up by other sources/examples.

Sustainable tourism

Checkpoints

1 **Benefits for the visitor**
The development of a quality tourism service.
Better relationships with the local community.
Closer involvement with and better understanding of both the people and the holiday destination.

Benefits to the tourism industry
Economic benefits for operators – e.g. reducing energy by installing efficient insulation.
Enhanced appeal of Northern Ireland for visitors from market areas that have ecologically-aware consumers.
Opportunities for the development and promotion of environment-friendly activity tourism e.g. conservation holidays.

Benefits to the environment
Safeguarding the resource for the benefit of future generations.
The protection and enhancement of the special landscapes which form much of Northern Ireland's appeal to visitors.

Benefits to the community
Opportunities for community involvement in tourism and the creation of a better climate for development.
Supporting the local economy and local services e.g. helping to support local transport systems in rural areas. Creation of new business opportunities.

2 The four distinctive characteristics are:
1 it covers a large area, over 2200 km², and involves over 40,000 people
2 it is managed by local people since the government believes that it will only work if it involves and benefits the local community
3 it is based on landuse zoning
4 part of the funding comes from the tourist trade.

3 Fees paid by trekkers passing through the Annapurnas fund ACAP. It has devised a multitude of simple fuel-saving devices, including solar-heated showers, and runs health-related workshops. These workshops include hygiene (the necessity of boiling water or digging latrines) and using flues to prevent buildings filling up with smoke and contributing to one of the biggest health problems, tuberculosis.

Exam practice

The Annapurna Conservation Area is visited by over 80,000 people a year. Over 700 tea shops and lodges have been built. Today, the local people are learning to maintain a high level of control over their resources and their future by building an endowment fund from entry fees to the area and working together on education, community development, biodiversity, and energy conservation projects. However, environmental and social consequences have been severe.

The concern of mountain people is that they will lose control over their culture, their economy, and their environment as tourism develops. Merely restricting tourism cannot be the solution to this imbalance, because people's desire to see new places will not disappear. Instead, mountain communities must achieve greater control over the rate at which tourism grows.

Except in the more popular routes, trails are in poor condition; acute deforestation has led to frequent landslides; boarding and lodging facilities are poor; and clean drinking water is not always available. Yet, because tourism is the country's second highest foreign exchange earner (after carpet exports), contributing nearly 4% of gross domestic product, the government is keen to attract more tourists. Nevertheless, after initial scepticism, local people have come to realize that they have a lot to gain from the situation.

Health and welfare

Checkpoints

1 HIV is spread among a population. After 5–10 years many AIDS cases emerge. Many workers become ill and die. The economy suffers. Lack of money to treat and prevent spread of the disease means the disease continues to spread. This leads to an increased mortality rate. The economy continues to decline and development is held back.

2 Cardiovascular diseases (e.g. heart attacks, strokes) are increasing in MEDCs. These are linked to increasingly sedentary occupations, rich diets and a lack of exercise. Deaths from cancers also contribute greatly to mortality rates, with other significant illnesses relating to use of alcohol and tobacco. These diseases can be reduced by a change in personal behaviour.

These diseases have costs: health education to aid prevention; complex and on-going treatment (high cost drugs); counselling and support services (e.g. Macmillan nurses); research and development – investment essential (e.g. Cancer Research UK). Despite medical research, the hope of a cure for diseases such as Parkinsons' disease, multiple sclerosis, and many cancers is small. These are degenerative diseases where the suffer experiences a gradual and distressing decline in quality of life.

Exam practice

1 This is a straightforward question, but make sure you only write about transmittable diseases rather than ones acquired in other ways.

Describe in turn the different means of transmission, i.e. air-borne, water-borne, etc. Make sure you give examples to support each one. Using a flow diagram like the one for cholera can show a sequence clearly.

Use terminology like incubation and vectors to show your knowledge.

It is worthwhile concluding with a point about how some diseases are spread due to poor living conditions (e.g. TB, cholera) as compared with those spread by certain social behaviour (e.g. HIV).

2 Begin by stating that tackling disease needs many strategies. It is useful to distinguish between preventative measures (such as vaccinations) and treatment measures (such as antibiotics and surgery). Make the key point that all measures have a cost attached. Outline some of the strategies that can be undertaken but point out that whilst the solutions are often relatively straightforward (e.g. pit latrines), so many of them are needed that it is difficult to provide the required changes quickly. This means that although some areas will be improved, diseases will

continue to spread in other areas. (You can usefully bring in knowledge from other topics relating to development and housing conditions here.)

The fact that many schemes need to be repeated, e.g. spraying of malaria-infested areas, may limit their effectiveness on cost grounds.

The potential spread of HIV in LEDCs could be linked to poverty, as condoms are not widely available. In conclusion, show that by tackling problems from a variety of angles the overall impact of disease can be reduced. However, unless capital (money) is made available from other sources, many LEDCs still find it difficult to implement comprehensive change as they are underdeveloped and their populations are still growing rapidly.

3 Begin by stating that disease has a cost to a society both in terms of responding to the diseases and in the resultant underproductive workforce; this gives a useful structure. Outline the ways in which money is spent on strategies to reduce disease such as vaccinations, public health schemes, antibiotics, etc; remember to talk about preventing and treating disease. Point out that both rich and poor countries have to pay, because different diseases emerge as countries develop.

The second section should focus on how disease can hold back the development of a country not just by the costs outlined above but due to the reduction in workforce productivity as a result of illnesses. The possible effects of an AIDS epidemic could usefully be given here.

Do not always assume that the answer is one-sided! In this case the main effects of disease are clearly negative but it could also be argued that some people might benefit from disease. These include pharmaceutical companies and healthcare equipment manufacturers. Don't be afraid to give a different perspective; this shows deeper thought.

Waste disposal

Checkpoints

1 The three Rs of waste disposal are: recycling, re-use and reduction. Give examples of what these things entail.

2 An example of substitution is the use of renewable resources, such as solar power rather than non-renewable resources, which reduces the amount of waste produced.

3 Landfill is cheap and relatively easy to do – it involves digging a hole, filling it with waste and then covering it.

Exam practice

The question 'To what extent' requires an answer – either it is or it is not. On the one hand, the recycling of computers is a good thing. Copper, gold and silver are recycled and this means that less mining is required and that resources are being used a second time – or more. However, the amount of material that is recycled is very small and the means used to recover the material is dangerous. For example, PCs are placed in baths of acid to strip metals from the circuit boards, a process highly damaging to the environment and to the workers who carry it out.

A third of the UK's waste plastic and paper (200,000 tonnes of plastic rubbish and 500,000 tonnes of paper) is exported to China each year. Low wages and a large workforce mean that this waste can be sorted much more cheaply in China, despite the distance it has to be transported. The transport of this material is harmful to the environment (greenhouse gas emissions).

In conclusion, countries such as China, that are importing waste, are increasingly aware that this is not 'responsible recycling' and that countries are exporting their pollution to them. They need to impose and enforce stricter laws on what types of waste can be imported.

Revision checklist
Sustainability

By the end of this chapter you should be able to:

1	Describe the principles of sustainability.	Confident	Not confident. **Revise** pages 150–151
2	Describe and account for the main characteristics of sustainable environments.	Confident	Not confident. **Revise** pages 152–153
3	Describe and explain the main characteristics of sustainable food supply.	Confident	Not confident. **Revise** pages 154–155
4	Describe the and explain the main characteristics of sustainable cities.	Confident	Not confident. **Revise** pages 156–157
5	Illustrate the issues connected to sustainable water supplies.	Confident	Not confident. **Revise** pages 158–159
6	Illustrate the issues connected to energy.	Confident	Not confident. **Revise** pages 160–161
7	Explain the issues related to sustainable energy supplies.	Confident	Not confident. **Revise** pages 162–163
8	Describe and explain the main features of sustainable tourism.	Confident	Not confident. **Revise** pages 164–165
9	Describe and account for the issues related to health and disease.	Confident	Not confident. **Revise** pages 166–167
10	Describe and account for the issues related to waste disposal.	Confident	Not confident. **Revise** pages 168–169

Global geography

The use of examples at a variety of scales is an important tool for geographers. One of the most important scales for human geographers is the national scale, where examples are drawn from countries. Countries have traditionally been grouped into More Economically Developed Countries (MEDCs), Less Economically Developed Countries (LEDCs) and Newly Industrialising Countries (NICs) or Recently Industrialising Countries (RICs). Most specifications now also require candidates to have a detailed understanding of one of the world's fastest growing and largest 'superpower' economies, India or China – and even where a specification doesn't explicitly include one of these themes, they provide excellent case studies and examples for almost every topic within human geography.

Exam themes

The themes will depend very much on the topic being studied or answered. The key themes are the economic, social and environmental changes that are occurring in the country. The themes developed here could be developed in other exemplar countries, but the exemplar countries we use here are:

- Superpower geography – China
- Superpower geography – India
- Newly industrialising countries (NICs).

Topic checklist

	Edexcel		AQA		OCR		WJEC	
	AS	A2	AS	A2	AS	A2	AS	A2
Superpower geography: China	O	●			O	●		●
Superpower geography: India	O	●			O	●		●
Newly-industrialising countries (NICs)	O	●	O	●	O	●		●

Superpower geography: China

Action point

Draw basic sketch maps of China showing population density, climate and the main economic regions (the Yangtze Delta (around Shanghai), the Pearl River Delta (around Guanzhou and Hong Kong) and the Beijing region).

China has one of the world's longest histories and oldest cultures. For much of its history it has been a leading world power, but for the last two centuries it has suffered social and economic decline, revolution and turmoil. It is the world's third largest country by area, but the largest by population. Since the end of the Cultural Revolution (1978), its economy has grown rapidly under strong government direction encouraging individual enterprise, and it is one of the four fast-developing economies, known as the BRIC economies (Brazil, Russia, India and China). It is now the world's fourth largest economy, growing at over 10% per year, and the world's second largest exporter of goods and services.

China's demography

Checkpoint 1

What are the main population problems facing the Chinese government?

China is the world's most populous country, with a population estimated at 1,322 million in 2007. About 58% of the population lives in rural areas, compared to 83% in 1978. The major cities are large and are growing rapidly by rural-urban migration – official populations are Shanghai (16.3 million), Beijing (10.3m), Shenzhen (11.8 m), and Guanzhou (7 m), but the wider urban area of each city is much larger.

Key issues facing the Chinese government are:

→ **Continuing population growth**. Even though the rate of population increase has declined due to policies such as the 'one child' policy, the population continues to grow, and will reach 1,440 million by 2025.

→ **Gender imbalance**. The desire for male children has led to abortion and infanticide of females, so the population ratio among under 5s is 117 boys: 100 girls.

→ **Ageing population**. The one child policy has been successful, but will result in an increasingly ageing population over the next three decades. In 2004, 10% of the population was over 60 years old; by 2040 33% will be over 60. This is known as the 4-2-1 problem (4 grandparents and 2 parents, supported by one child).

Action point

Use the Internet to find details of the growth and development of the Pudong suburb of Shanghai.

→ **Rural poverty**. This is a major problem, as most of the economic growth has been in the cities. Migrants from rural areas leave the poorest in the rural areas.

→ **Urbanisation**. In 1990, 222 million people lived in China's cities, but China's urban population is now 540 million. Urbanisation has developed alongside industrialisation, and while many urban workers are wealthier than their rural relations, urban unemployment and health and social issues are major problems.

Shanghai, for example, has a population of 16.3 million, including 3 million migrant workers, and will grow to 20 million by 2020. Its growth is predicted to produce a 'megalopolis' in the Yangtze Delta, where the total population in 225 cities is already 193 million.

China's economy

In the last decade, China's economy has been one of the fastest growing in the world (11.9% per annum). The Chinese economy now consumes 33% of the world's steel and 50% of the world's concrete.

Agriculture is an important sector, as 60% of the labour force works on the land, but they only contribute 12% to GDP. Productivity grew after the end of collectivisation in 1984. Most agriculture is in the fertile eastern coastal areas and the main river valley areas. Key crops are wheat in the north and rice in the south.

Industrial development has been the main feature of China's economic success. The emphasis has been on light manufacturing industry, high-tech industry and the automotive industries, promoted by encouraging individual enterprise and attracting foreign investment to Special Economic Zones (SEZs), e.g. in Shenzhen. The urban areas are the main industrial centres, in particular Shanghai, Beijing, Chongqing, Shenzhen and Guanzhou. The growth has included large investment in transport and infrastructure.

Services employ 20% of the labour force, but provide 40% of GDP. Tourism is of growing economic importance and China now has more tourists than any other country.

China's environment

China faces many environmental challenges:

→ **Natural hazards**. Typhoons are a threat to the south-east coast (from Shanghai to Hong Kong). Earthquakes are a major natural hazard (e.g. the Szechuan earthquake of 2008).
→ **Watershed clearance and flooding**. Clearance of forest from mountain areas and poor farming techniques lead to soil erosion and flooding, especially in the Yellow (Huang He) River, which drains easily-erodible loess soils, and in the Yangtze delta. Desertification is a problem in many semi-arid areas.
→ **Pollution**. Industrialisation has caused serious pollution problems, due to waste disposal and the use of coal-fired power stations. One new power station opens every week in China. Air pollution is seen as the cause of 400,000 deaths per year. Two-thirds of China's rivers and lakes are polluted and access to clean water is a problem for most Chinese cities. A 1998 WHO report identified seven of the world's ten most polluted cities as being in China.
→ **Biodiversity issues**. Pollution and habitat loss have threatened species (e.g. Giant Panda, Yangtze River Dolphin). Industrial projects threaten habitats, e.g. the Three Gorges Dam Project. Government is now recognising the need to protect threatened habitats and 7.8% of the land area is now protected (e.g. as national parks or reserves).

Exam practice
answers: page 182

1 With reference to China, outline the social, economic and environmental threats to its development. (12 mins)
2 What are the major population issues facing China in the 21st century? (12 minutes)

Checkpoint 2

What have been the main reasons for China's rapid economic growth since 1980?

Examiner's secrets

You can use the examples you learn from China to illustrate almost any of the themes in your A2 or AS geography examinations. This also shows that you are able to use your geographical knowledge in a range of contexts, and will enable you to pick up extra marks.

Action point

Use the Internet to identify the details of the Szechuan earthquake (2008).

Checkpoint 3

What are the main environmental issues facing China?

Grade booster

For each of the environmental issues facing China, learn a specific case study example to illustrate the causes and also the Chinese government's strategies for dealing with each of them.

Superpower geography: India

Action point

Draw basic sketch maps of India showing population density and the main economic regions. You should be able to illustrate many of the themes of your course with examples from an LEDC such as India.

India is a former British colony that became independent in 1947 and was partitioned from Pakistan (including modern Bangladesh). It is the world's seventh largest country by area, but its second largest by population. It is the world's twelfth largest economy and one of the four fast-developing economies known as the BRIC economies (Brazil, Russia, India and China).

India's demography

A key characteristic of India is its large and rapidly growing population, which has risen from 913 million in 1994 to 1,027 million in 2001, and to an estimated 1,132 million in 2008. By 2025 the population will be 1,440 million and India will become the world's most populous country. About 70% of the population lives in rural areas. The major cities are large, and are growing rapidly through rural-urban migration: Mumbai (formerly Bombay – 13.6 million), Delhi (11.9m), Kolkata (formerly Calcutta – 5.2 m), Chennai (formerly Madras – 5.2 m), and Bengaluru (formerly Bangalore – 4.5m). The annual population growth rate is 1.4%, with an average of 3.7 children per woman.

Key issues facing the Indian government are:

→ **Rapid (but declining) population growth.** Even though the rate of population increase is now slower, the sheer numbers drive rapid growth.

→ **Health.** Life expectancy and health levels are rising, but health care in rural areas and urban slums is a major challenge, with one doctor per 2,400 people, under-5 mortality rates of 119 per 1,000, and maternal mortality of 437 per 100,000 live births.

→ **Income inequality and wealth distribution.** Rural poverty is a major problem, but the urban middle classes have growing incomes and standards of living.

Checkpoint 1

What is a bustee?

→ **Migration to the cities and urbanisation.** The growth of bustees, urban unemployment and health and social issues, including water supply and sanitation, are major issues. Mumbai, for example, is the third largest megacity in the world – its urban area is 18.8 million people. It grows by 300 people per day through migration, and has doubled in size since 1985. Slums are the home to 54% of its population. Major problems in Mumbai are overcrowding, disease, flooding (due to the riverside locations of bustees), transport and provision of basic amenities (water and electricity).

Grade booster

Practise drawing quick sketch maps of India with annotations of the main issues or locations that you need to refer to. This will save time in the exam and also gain you extra marks.

India's economy

In the last decade India's economy has been one of the fastest growing in the world (9.4% per annum). 60% of the labour force work in agriculture and village life is still the basis of the economy. Many are landless or have very small landholdings (under 2 ha); many sharecroppers own no land and pay rent from crops, which results in low investment and poor yields. The Green Revolution has helped to improve yields. Key crops are wheat, rice and cotton.

Checkpoint 2

What is the Green Revolution and what advantages and disadvantages has it brought to countries like India?

Industry employs 12% of the population, with some of the world's largest corporations (e.g. Tata). This ranges from heavy industry (e.g. iron and steel in the Damodar Valley) and chemicals, to cars, food processing and textiles. IT and computing is a growing sector – Bengaluru is home to major software producers.

Services employ 28% of the labour force, but provide 54% of GDP – there is large government employment, but a large proportion of this is 'offshore' services in finance, banking and IT. Relatively low salaries, combined with good English language and education levels, means that telephone services for UK or US companies is an important sector. Tourism is a small but growing employer.

Checkpoint 3

Why do many British companies have telephone helplines that are serviced from India?

India's environment

India faces many environmental challenges:

→ **Weather hazards**. Typhoons are a threat in the Bay of Bengal (e.g. in Orissa 1999 and 2001) and heavy monsoon rainfall brings flooding.
→ **Watershed clearance and flooding**. Clearance of forest from mountain areas leads to soil erosion and flooding, e.g. in the Ganges delta.
→ **Pollution**. Rapid industrialisation leads to pollution problems and disasters (e.g. Bhopal 1984). Urban pollution is due to rapid rises in car ownership and the use of coal and lignite-fired power stations.
→ **Threats to biodiversity**. India has 500 wildlife sanctuaries and 13 biosphere reserves, but many species (e.g. Bengal tiger) and environments are under threat.

Action point

Use the Internet to identify the details of a recent weather disaster in India.

Action point

Use the Internet to find the details of Project Tiger.

Examiner's secrets

These notes are intended to give you no more than a few key themes in the study of India as a superpower. Your classroom notes should provide you with more. Keep your own file of newspaper cuttings that might provide you with up-to-date information. The *Financial Times* has an annual supplement on India: check it out on www.ft.com.

Exam practice   answers: page 182

1   Outline the social, economic and environmental threats to India's development. (12 mins)
2   Study the model of the interlocked cycles of poverty shown below. Explain how this model relates to an LEDC that you have studied. (15 mins)

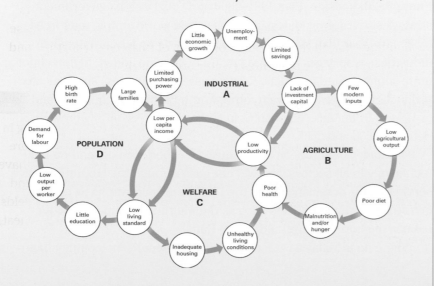

Newly-industrialising countries (NICs)

Newly Industrialising Countries or NICs (sometimes called Recently Industrialising Countries or RICs) are relatively far along the path to industrialisation or a modern economy. They are the locations for much modern industrial production, particularly in the fields of electronics and IT equipment.

Rostow's model

Economic development is often described using Rostow's model. This shows how an economy changes from a traditional society to a high mass consumption society.

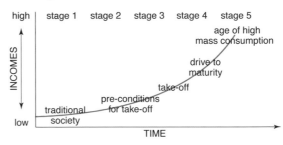

Malaysia's development to an NIC: Rostow revised

Malaysia, in south-east Asia, is an example of an NIC, and its development can be seen as occurring through a number of phases.

Phase 1: Early industrialisation. The 1956 economy is geared to the export of rubber, palm oil and timber. Tin and oil are the main mineral exports. **Import substitution** – the production of goods needed by the population, e.g. clothing, to avoid expensive imports.

Phase 2a: The new economic policy 1971–1990. Oil exports fuel growth. Government initiatives and central plans enable rapid growth of manufacturing + 26%. Creation of industrial estates; investment incentives, especially for Japan. In 1970, 60% of manufacturing was foreign owned. The process was government-led and not *laissez-faire* – a major departure from Rostow, market-led investment.

Phase 2b: Push to heavy industrialisation and export-oriented electronics firms through state planning. Cars, steel, motorcycles, oil and gas. Tourism and education developing.

Phase 3: 2020 vision. Plan 1991–2000 and '2020 vision': government strategies to leapfrog Malaysia to developed world status, reinforced by 9th Five Year Plan starting in 2006. Emphasis on labour-intensive industries – shortages of skilled labour, high value added, high-technology, more sophisticated products. Routine assembly and less sophisticated production processes move to other South East Asian countries and China. In 1994, 360 Japanese companies in Malaysia. The economy grew by 50% between 2000 and 2005.

The stages of European post-industrial experience depend on high-technology in all sectors, and will not pass through the labour-intensive service sector dominance as in the UK. There is an emphasis on state investment in huge projects such as the Multimedia Super Corridor, Labuan offshore banking, and Bakun Dam.

Where does Malaysia stand on Rostow's model? The country has progressed towards industrial status, and is described as an NIC by the World Bank. Malaysia prefers the title **newly industrialised economy (NIE)** whereas media describe it as an 'Asian Tiger', to convey an image of aggressive economic behaviour.

The tiger growth model

The rapidly growing economies of Asia in the 1980s and 1990s (e.g. Taiwan, South Korea and Singapore) were described as tiger economies, and the model below shows their typcial pattern of growth.

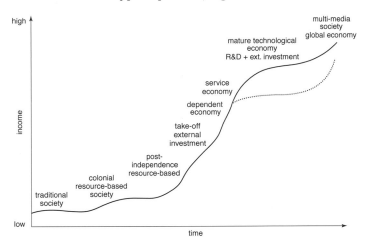

NICs as growth locations for manufacturing

Why do NIC governments see manufacturing as important to economic growth?

→ Countries wished to substitute manufacturing for imports.
→ Able to attract foreign investment, especially from the regional leaders such as Japan in South East Asia.
→ Governmental creation of right conditions such as **EPZs**.
→ State enabled very large firms with a range of products – **Chaebol** in South Korea or **Zaiobatsu** in Japan.
→ Emphasis on consumer products that have high demand world-wide.
→ Strict labour laws, e.g. it is illegal to strike in Taiwan.
→ For the most successful, an ability to attract new investments in the tertiary sector and to invest in the region.
→ TNCs have tended to favour NICs, which has led to a clustering of similar activities, e.g. in Penang, Malaysia.

The social impacts of growth include a greater divergence in earnings, employment of women, use of ex-patriates, Westernisation of life, multiplier effect on, for example, purchasing patterns, growth in training related to NICs.

The jargon

EPZ is Economic Priority Zone. Many countries, especially in Asia, have a range of acronyms and abbreviations associated with all forms of development – *FIZ* (Free Investment Zone) and *FTZ* (Free Trade Zone) are two others used in Malaysia.

Action point

If you are using another NIC, draw up your own examples of each of these points. Also, have some detailed knowledge of the effects of development on people and the local environment.

Grade booster

Showing you have up to date knowledge of NICs will impress the examiner – use very recent examples or data from the last two years to illustrate your answers. You can get this from the web or from newspapers.

Exam practice   answers: page 182

1 Explain why Rostow's model of development is less appropriate today. (8 mins)

2 To what extent have NICs begun to move to MEDC status? (12 mins)

Answers
Global geography

Superpower geography: China

Checkpoints

1 High annual increase in population, but slowing of the birth rate leading to ageing population. Gender imbalance. Rural poverty, rural-urban migration and urbanisation.
2 Encouraging individual enterprise and private companies. Attracting foreign investment, e.g. by setting up SEZs. Relatively low wage levels. Expertise in light manufacturing industry and providing for global export markets.
3 Natural environmental issues, e.g. typhoons, earthquakes. Pollution and habitat loss due to industrialisation and urbanisation.

Exam practice

1 Start by defining social, economic and environmental threats. Social threats include rural poverty, issues from rapid urbanisation, ageing population, gender imbalance, plus issues over individual human rights. Economic threats include the dependence on global trade, so susceptible to global recession, competition from other BRICs and NICs, e.g. dependence on imported raw materials, and growing wage levels, which may remove one of China's main economic advantages. Environmental threats include urban pollution and political pressure from other countries to deal with national environmental issues and to contribute to tackling global issues such as climate change.
2 Describe the growth of China's population and indicate its position as the most populous country. Then discuss each of the main issues: continuing population growth; impact of 'one child' policy in rural and urban areas, gender imbalance, and ageing population with its potential economic impacts. Describe the future predicted population growth, with stabilization after 2025. Discuss the social and economic issues associated with urbanisation and rural poverty.

Superpower geography: India

Checkpoints

1 Bustee is an Indian name for a squatter settlement.
2 The Green Revolution is the development of new high-producing hybrid plant crops to increase food production. Advantages include increasing food production. Disadvantages include the cost of buying seed, leading to debt, the need for higher use of fertilizers and the risks of dependence on a narrow range of crops which may be more susceptible to disease.
3 India provides well-educated, English-speaking staff at lower labour cost than in the UK, and global telecommunication costs are now quite low.

Exam practice

1 *Social* refers to problems connected with people, e.g. in India there are huge numbers living in squatter settlements, as well as a caste system that divides society. There is the issue of population growth and control, as well as health issues, including AIDS. *Economic* refers to many of the problems that demography generates, e.g. coping with the housing, health, education and employment needs of such a large population.
Environmental problems can be physical, e.g. failure of the monsoon rains, or human, e.g. pollution.
2 Take each circle in turn, A–D, then find its start point and explain the sequence, e.g. Population – the high birth rate in rural India leads to large families of about five children. Because these families have low earnings due to lack of employment possibilities, living standards are low and only 23% of children attend secondary school, etc. You need to explain the key overlaps of the circles.

Newly-industrialising countries (NICs)

Exam practice

1 All models are a product of the time they were written. Rostow assumed a standard, capitalist route to modernisation. Socialist states took a different route. NICs took a different route as no need to pass through early industrial phases. New economic activities and role of TNCs.
2 You will need statistics to be able to demonstrate that the country is similar to an MEDC. Some aspects might lag behind, especially social indicators. The answer is generally that NICs are closing on MEDCs in an economic sense but lagging in terms of social development. Because the question asks you to evaluate, you will need to have some contrasting NICs.

Revision checklist
Global geography

1	Describe and explain the demographic issues facing China.	Confident	Not confident **Revise** pages 176-177
2	Describe and explain the economic issues facing China.	Confident	Not confident **Revise** pages 176-177
3	Describe and explain the environmental issues facing China.	Confident	Not confident **Revise** pages 176-177
4	Describe and explain the demographic issues facing India.	Confident	Not confident **Revise** pages 178-179
5	Describe and explain the economic issues facing India.	Confident	Not confident **Revise** pages 178-179
6	Describe and explain the environmental issues facing India.	Confident	Not confident **Revise** pages 178-179
7	Explain and illustrate, using examples, the characteristics of Newly-industrialising countries (NICs).	Confident	Not confident **Revise** pages 180-181
8	Use Malaysia as a case study of an NIC.	Confident	Not confident **Revise** pages 180-181
9	Describe and explain the main models of economic development, e.g. Rostow's model.	Confident	Not confident **Revise** pages 180-181
10	Use China and India as examples of the economic development in the BRIC countries.	Confident	Not confident **Revise** pages 176-179

Resources

Successful study of geography at AS and A2 level requires you to be aware of the main techniques for studying the subject and for using your knowledge and skills to best effect in the examinations. Many students who have a good knowledge of the subject matter of geography do not perform as well as might be expected because they make errors in the way they interpret or answer questions, or in how well they demonstrate general skills, such as using examples, and geographical skills such as drawing sketch maps.

Being successful also requires you to show understanding of how the parts of geography fit together. Although you study your course in units or modules, these are only convenient ways of dividing the subject – they are not separate subjects, and there are many overlaps between them and common ideas that run through them. This section looks at some of the key aids to studying that will help you towards success.

Exam board specifications

In order to organise your notes and revision you will need a copy of your exam board's syllabus specification. You can obtain a copy by writing to the board or by downloading the syllabus from the exam board's website.

AQA (Assessment and Qualifications Alliance)
Publications Department, Stag Hill House,
Guildford, Surrey GU2 5XJ
www.aqa.org.uk

EDEXCEL
190 High Holborn, London WC1V 7BH
www.edexcel.org.uk

OCR (Oxford, Cambridge and Royal Society of Arts)
1 Hills Road, Cambridge CB2 1GG
www.ocr.org.uk

WJEC (Welsh Joint Education Committee)
245 Western Avenue, Cardiff CF5 2YX
www.wjec.co.uk

Topic checklist

	Edexcel		AQA		OCR		WJEC	
	AS	A2	AS	A2	AS	A2	AS	A2
Approaching and using fieldwork	○	●	○	●	○	●	○	●
Illustrating answers	○	●	○	●	○	●	○	●
Synoptic assessment and use of pre-release resources	○	●	○	●	○	●	○	●
Making use of projects and enquiries	○	●	○	●	○	●	○	●
Examination question words	○	●	○	●	○	●	○	●
Effective revision notes	○	●	○	●	○	●	○	●

Approaching and using fieldwork

"When we try to pick out anything by itself, we find it hitched to everything in the universe."

John Muir, My First Summer in the Sierra, 1869

Geography is about the world we live in and its complex interrelationships. To study that world we need to examine it first hand whenever possible. Therefore your AS and A2 studies will expect you to show that you have studied some of that world at first hand, through fieldwork.

Not all fieldwork has to be led by your tutor. Good tutors will take you into the field whenever possible and, if you are particularly fortunate, you may have spent a period of time away in a different environment. The examples that you see in the field, e.g. types of agriculture, an out-of-town retail park, an industrial estate, the impact of a landslide, should be used to support your written work in examinations. However, you should be able to use your own eyes, while journeying about your home area and while on holiday, to see other examples that you can use. These examples should be added to your own notes in the relevant place.

Primary and secondary data

Primary data will come from fieldwork. Field measurements of stream discharge, maps of retailing in a city centre or quadrant sampling of vegetation are all examples of primary data collection.

Good fieldwork investigations, though, will also use **secondary data** because you do not have the time or money to enable you to visit everywhere. Secondary data will also often help you to decide the best locations for fieldwork. One of the standard secondary data sources will be the textbook that covers most of your specification. Other good sources of information are journals such as *Geography Review* and *Geographical*. *National Geographic* also has some relevant case studies such as the one on Hurricane Mitch in November 1999.

In this electronic age there is a whole range of secondary sources, some of which can almost be described as primary sources. The **Internet** is an excellent source if it is used carefully. For instance, the US Geological Service website had details on the Turkish earthquakes shortly after they happened. Data on the Smoke Haze over South East Asia has been readily available on the Singapore Met Service site. The advantage of the web is that it can be up to date. Nevertheless be wary; some sites – especially those related to economic and social geography – may be biased because they are produced by pressure groups or companies with a specific marketing objective. Always see if you can find a site with an alternative view if you think that the opinions on one are too biased.

Fieldwork and geographical enquiry

Good fieldwork in geography uses an 'enquiry' approach. Enquiry has a specific route that you can follow, no matter what the topic:

→ *Describe* and *define* a phenomenon or issue – say **what** it is and **where** it can be found.
→ *Explain* the phenomenon or say **how** it came about, and *predict the outcomes* or suggest **what will happen if** something occurs.
→ Finally, one can *evaluate* and *prescribe* or say **how one should proceed** and how this solution or answer compares with alternative solutions.

For example, a fieldwork study of a stretch of coast might *describe* the coastal landforms of the mouths of Chichester Harbour and Langstone Harbour, *explain* how the spits were formed, *predict the outcomes* if people can roam all over the dunes at West Wittering, *evaluate* the measures to conserve the dunes and spits and, finally, *prescribe* the most satisfactory conservation solution on the basis of our knowledge of spits and dunes.

Planning fieldwork

Field work always needs careful planning. You will need to:

→ Be clear about the aims of your fieldwork – what are you trying to find out?
→ Identify the key question for the work or the hypothesis you are going to test.
→ Do lots of background reading on the topic, on the location you will be visiting and on the methods you will be using (e.g. sampling methods or statistical tests).
→ Undertake a careful risk assessment.
→ Pilot any questionnaires you wish to use and practise any fieldwork techniques (e.g. river discharge measurements) BEFORE you start the real fieldwork.
→ Collect the data as carefully as you can in the field. Remember to take digital photographs where they will be helpful.
→ Record the data in tables, charts, diagrams, etc.
→ Analyse the data using quantitative methods (e.g. statistics such as Chi-squared tests) or qualitative methods (e.g. analysing interviews).
→ Be clear what your conclusions are at the end of the fieldwork.
→ Evaluate the methods you have used so that you can provide a critique of what you have done.
→ Present the findings carefully as a fieldwork report.

Action point

List all the fieldwork you have done during your geography courses, including work you did at GCSE level. Think which topics in your current course the fieldwork links to.

Action point

Plan a fieldwork enquiry on river floods, or a derelict site in the inner city, or migration into a city.

Watch out!

Ensure for your safety that you always do a risk assessment before undertaking fieldwork.

Examiner's secrets

There are always valuable marks given for the Evaluation section of a fieldwork report – and many students score badly on this. Remember, this is the place where you can show your skills at knowing what makes good and bad fieldwork.

Illustrating answers

Your grade in geography will depend in part on how well you can show that you know real examples of the ideas and issues that you are writing about. Geography is the study of the world around us, so you've got to show that you know case studies, and can both describe and illustrate them, for example with maps and diagrams.

Examples and case studies

Throughout your course in geography you should always include examples and case studies in your notes and in your written work.

Action point

Check your particular exam specification to find out what case studies it asks for in each unit. Check your own notes to be sure you have these case studies.

→ An example is simply the name and a few brief details of a real place or situation, e.g. an example of an LEDC city is Mexico City, with a population of about 22.5 million and a growth rate of 5% per year.

→ A case study is a much more detailed example that you can use to discuss a range of issues relating to the topic.

Building a range of examples

You may be asked to know examples in a range of different scales, or from countries at different levels of economic development, and for each of the topics you have studied. The table below lists scale types and examples in some topics within geography.

Action point

Draw up a table like the one here for each of the topics you have studied. If there are gaps, then check the specification to see if you need an example for that gap – if so, find one in a textbook and make notes on it.

Study focus	SMALL/LOCAL SCALE (S)	REGIONAL (R)	NATIONAL (N)	INTERNATIONAL (I)	GLOBAL (G)
Rivers	Small-scale catchment, e.g. River Meon Small-scale ecosystem, e.g. a wood or heathland in the New Forest	Large drainage basin, e.g. the Thames		Rivers with varied regimes, e.g. Mississippi/Rhine Impacts of rivers on people, e.g. floods Conflicts and consequences Management solutions	Hydrological cycle
Coasts	Landforms and processes on small section of coastline, e.g. Barton-on-Sea Sand dune and one other ecosystem, e.g. West Wittering Impacts of coastal change, e.g. Holderness	Extended coastline Coastal land uses and conflicts Management strategies	Coastal management strategies, e.g. The Netherlands	Contrasting examples of conflict, e.g. Mediterranean	Coastal environments Distribution of examples of coral reefs Sea-level changes
Rural	Changing villages e.g. in the Derbyshire Peaks Rural planning issues, e.g. in the Surrey greenbelt	Rural–urban continuum, e.g. the Vale of York Rural planning issues, e.g. in Hampshire Contrasts in rural areas, e.g. Cumbria Conflict and change in rural environments, e.g. a National Park		LEDC/MEDC contrasts, e.g. Ghana/France Two contrasting rural environments Future development strategies	Changing importance of rural/urban Global ruralisation
Urban	CBD changes, e.g. Birmingham Rural–urban fringe conflicts Managing urban environments, e.g. new towns Sustainable urban projects	Land use zoning in a large urban area, e.g. Bristol Managing urban environments, e.g. Bath	Cycle of urbanisation in one country, e.g. UK	LEDC/MEDC contrasts, e.g. London/Mumbai Contrasting world cities, e.g. Paris/Tokyo Quality of life Sustainable initiatives	Global urbanisation Global growth of megacities, e.g. Mexico City/Beijing

Providing examples in exams

Whatever the form of assessment (traditional exam, timed essay, data response questions, etc.) you will be expected to give real examples in your answers. You may be asked to do this in various ways.

No instructions

Most questions do not ask you for examples, but you must still provide them. Illustrate each idea you present with a named example or a brief case study. Examples from your own fieldwork are a good idea.

With reference to specific examples/case studies

This is asking you to build your answer around one or two detailed case studies, e.g. a case study of managing visitor pressure at Tarn Hows in the English Lake District. Check the number of examples or case studies you are being asked for. Within the case study you will be expected to know the detail, i.e. names of places, production figures, dates of key events, names of organisations involved, etc.

Action point

Look at the questions in past exam papers or specimen papers. Against each one, note down which examples and case studies you would use.

Locating examples

Whether your examples are brief or detailed as in a case study, you must locate the example carefully. Say exactly where it is and, if possible, draw a simple sketch map. For a case study a sketch map is almost essential. Where you want to draw a map of the case study area *and* locate it in a region, country or continent, draw a small inset sketch map next to your main map to show where it is.

Action point

Get into the habit of summarising ideas or facts in a table rather than as text. It is quicker and easier than writing it all out.

Important note!

Don't let your case studies and examples be a straitjacket for you.

→ Always pick out the important and relevant details to include. An exam question will ask you to use ideas from a case study. *Don't* just write down everything you know about the case study.

→ Mix and match the case studies and examples. Use some ideas from one case study together with some from another if necessary. Remember – because you learned a case study in a unit entitled 'urbanisation' doesn't mean you can't use it to answer a question on sustainable development if it is relevant and helpful.

Illustrating your examples

Sketches, maps, diagrams and charts are important in geography because:

→ they provide a visual image of data
→ they provide a useful summary of data
→ they demonstrate you can use key geographical skills.

In all exam mark schemes some of the marks are for illustrative material, so make sure you pick up those marks – even where the question doesn't specifically ask for illustrations.

Grade booster

For *every* case study you have, practise a simple sketch map that you can draw in a minute or two in the exam room. Remember, it doesn't need to be a work of art – it's a quick sketch map!

Examiner's secrets

Many candidates simply throw away the marks that may be gained for good sketch maps or diagrams by not bothering to include them.

Synoptic assessment and use of pre-release resources

Action point

Check your particular specification to find out how the synoptic assessment operates.

All A2 specifications include 'synoptic assessment' as part of the assessment towards your final grade. Synoptic assessment is designed to test how well you can:

1 understand the overall nature of geography as a subject
2 understand the links between the parts of geography
3 draw on that understanding to investigate new or different situations using your geographical knowledge, understanding and skills.

This is normally tested through one or more of the following:

→ synoptic questions in a written examination, typically within essay questions exploring geographical issues or problems, key geographical concepts or a particular geographical region
→ personal investigations/projects or fieldwork
→ decision-making exercises, problem-solving exercises or issues evaluation tasks.

Synoptic exam questions

These are questions based on particular ideas or themes that draw together many aspects of geography. They are normally related to one of two broad fields:

→ people–environment interactions
→ sustainable development.

E.g. 'With reference to a major water resource development project, explain how the exploitation of water resources can bring about physical, economic and social changes in the immediate environment.'

This question requires you to show knowledge from many parts of your course – water management and its links with hydrological processes (e.g. flooding), economic systems (e.g. water supply and farming) and social/political issues (e.g. the rights of local people v. industrial demands). You will also need to know a case study that shows these links, e.g. the Indus Valley Project in Pakistan.

Personal investigations

These require you to draw on your knowledge and skills to investigate an individual topic. This is described in the section on 'Approaching and using fieldwork' on pages 186–87.

Issues analysis/evaluation

Examiner's secrets

Check whether Issues analysis/evaluation is included in your particular specification. If so:
→ How is it structured?
→ How are marks awarded for the different parts of it?

Issues analysis/evaluation (formerly Decision Making Exercises) is an exercise in which you are asked to use your geographical knowledge, understanding and skills to analyse a particular problem, question or issue and to propose a solution or strategy for dealing with that issue. You will be provided with an Advanced Information Booklet a few weeks before the examination with a wide range of data, views and other resources to enable you to have a full understanding of the issue – however, you will not be given the exact examination tasks until the exam itself.

Examples of the sort of issues you might be asked to consider are:

→ choosing between alternative routes for development projects for an island
→ choosing a strategy for managing a popular visitor location
→ choosing the best site for a new power station.

An Issues analysis/evaluation can require you to:

→ Take on a role to examine the issue. This may be a role, for example, as a journalist, an academic, a planner, a local resident, a government official or chair of an independent enquiry. In 'playing this role' you will be expected to show that you can understand an issue from a different person's perspective.
→ Study, analyse and interpret a range of information sources about the issue involved. These may be maps (at any scale), diagrams, tables of data, reports, summaries of the views of a range of interested parties and technical information about the issue and the range of possible 'solutions'. This will test your ability to interpret geographical data and use appropriate geographical skills.
→ Show that you understand the nature of the issue by describing its causes and the social, economic and environmental factors that have an influence on it. In describing the nature of any issue these three headings (social, economic and environmental factors) provide a good framework for structuring your answer. To score high marks, you will be expected to show that you understand the links and interactions between these three groups of factors.
→ Show that you can identify the range of different viewpoints about the issue. You will be expected to be aware that people's views on an issue depend on:
    → their vested interests in the issue
    → their own attitudes and values.
→ Critically review and analyse the existing proposals, using a range of techniques, such as cost–benefit analysis or environmental impact analysis. You may need to produce diagrams, tables or sketch maps.
→ Choose a preferred option, or identify a solution yourself.
→ Justify your choice with reference to the evidence.

You will also need to present your findings in a logical, well-argued way, using good spelling and grammar and making suitable use of graphs, tables and sketch maps.

Using pre-release materials

These are provided so that you can have a good understanding of the issue and have plenty of time to read the resources and materials before the exam. Make good use of this time. Read and think about all the materials: what questions do they raise, what alternative views do they represent? Read your AS and A2 notes on all the topics that appear within the materials so you are fully familiar with the key ideas and concepts, and other examples.

Making use of projects and enquiries

Projects and enquiries are designed for you to show how you can pull together your geographical knowledge and skills to investigate a specific topic or question. In particular they will test your skills of geographical enquiry.

The jargon

Projects are given many titles depending on exactly what they ask you to do, e.g. *personal enquiry, environmental investigation, personal investigation, personal investigative study*. Check what your specification calls it, and read the 'rules' for it.

Planning the project

Spend some time planning the project so that you know your deadlines and what you must do to meet the exam requirements.

→ Work backwards from the date you must hand your work in.
→ Allow 2–3 weeks more than you think you need, to allow for unforeseen circumstances (illness, projects in other subjects, etc.).
→ Identify a data collection time over a holiday if possible.
→ Build in plenty of writing time – A2 projects may take 40–50 hours of writing time!

The figure below shows a timeline for an A2 project.

Action point

Which two topics that you have studied have you found most interesting? For each, write down a question you could study for a project.

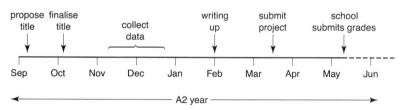

Cycle is applicable ***ONLY*** to those specifications with an A2 project

Choosing a topic

If you are free to choose a topic, choose one that genuinely interests you, so that you are motivated to do your best work. Choose a topic that is:

1. One you found interesting in a unit you have studied, *or*
2. About an issue in the local news, e.g. plans for a new road, *or*
3. About an issue at a national or international scale (e.g. the decline in beef farming), but look at the issue in your own locality.

Action point

Read a copy of your local newspaper. Identify three current news topics that you could use for a project.

Action point

Investigating a local angle on a global problem provides a good project. Try to identify some local issues in your area arising from poverty in LEDCs (e.g. sales of 'fair trade' products in local shops).

Key questions and hypotheses

Once you have chosen a general topic, you will need to identify a **key question** to ask, and/or a **hypothesis** to test. The key question may be the title of your project, e.g. 'What are the likely environmental impacts of proposed housing developments in Bursledon?' This will need to have several sub-questions following a route of geographical enquiry, e.g. *Why* is there pressure to build here? *Who* are the interested parties and what are their views? *What* alternatives have been put forward? What *should* be allowed to happen?

If you choose to test a hypothesis (e.g. 'Out-of-town shopping centres cause a decline in shopping services in the city centre'), the evidence should still be collected through a geographical enquiry route. The only difference is that at the end you will need to accept or reject the hypothesis.

Action point

For a project investigating issues about a proposed new housing estate on farmland, who might be the 'interested parties' you would need to contact?

Collecting background information

What is known already about this topic? Start by collecting information from a range of sources:

→ books, articles, newspaper reports, official reports, etc. from your geography department or library, or from your local library
→ the Internet
→ official bodies, e.g. the local council, charities, NGOs.

Action point

When contacting an organisation for information, be very precise about what you need, e.g. details of agricultural development projects in The Gambia, not 'anything on farming in LEDCs'. Always offer to pay for postage (or send a sae).

Collecting the data

Most projects require you to collect primary data, but you will also use secondary data as well. You will need to choose the most suitable research method to collect the data, and be able to justify your choice. You will also need to carry out the data collection in a reliable way. It is important not to collect too much data, or you will be overwhelmed by it – e.g. a sample of 50–100 questionnaires, or 10–15 interviews, or 10–20 soil pits/vegetation quadrants will be plenty.

Action point

Find out the difference between *random sampling* and *systematic sampling*.

Analysing the data

Data analysis has three stages:

1. Data presentation – use graphs, tables, maps, diagrams, overlays etc., but check you are using each method correctly. A good source book is D Richardson and P St John, *Methods of Presenting Fieldwork Data*, published by The Geographical Association.
2. Testing the data for relationships – either by using a statistical test (e.g. Spearman's Rank Test) or by examining it for apparent connections in the data. A good source is P St John and D Richardson, *Methods of Statistical Analysis of Fieldwork Data*, published by The Geographical Association.
3. Interpreting the data, to answer the key question or test the hypothesis.

Drawing conclusions

Your project must end with clear conclusions that:

→ summarise the main findings from your work
→ answer the key question, or accept/reject the hypothesis
→ critically review your study
→ make recommendations for managing the issue.

Examiner's secrets

A small-scale, highly focused project produces the best grades.
You will lose marks for poor spelling and grammar.

Writing up/presentation

Think carefully about:

→ the report's structure – you will need sections on 'aims', 'background', 'research methods', 'results', 'conclusions'
→ quality of presentation – use your ICT skills to the full
→ spelling, punctuation and grammar, i.e. Quality of Written Presentation.

Grade booster

Gain marks by using a range of presentational methods, e.g. graphs, tables, photographs.

Examination question words

Practise with sample exam papers by underlining the key command words in each question.
Think how you would answer this question: 'Discuss the issues involved in considering a planning application for a new opencast coal site'.

To make sure that your grade is as high as possible you must answer exam questions as carefully as possible. In particular, you must answer the question that *has* been asked, not the question that you *wish* had been asked. Understanding the precise meaning of the *command words* in exam questions is vital, as is checking that you are using the right type, scale and location of *examples*, covering the right geographical *concepts/ideas*, and that you are demonstrating the right geographical *skills*.

Key command words

Command words are the words that tell you what to do. The principal command words used are:

Describe ...

e.g. *'Describe the main features of a hurricane.'*
This is asking you to show that you know what is occurring in relation to a particular topic. You will need to show that you know **where** in general this occurs (say *exactly* where hurricanes occur), **what happens** (describe the main weather features of the hurricane), and the main **effects** of what is happening (e.g. wind damage, floods, disruption, deaths).

Examiner's secrets

Geographical enquiry questions provide a useful checklist for what you should be trying to answer in an exam question.
(Describing) What? Where? When? Who? What effect?
(Explaining) How? Why?
(Discussing) above + What will? How should?

Explain ...

e.g. *'Explain the causes and consequences of rural–urban migration in LEDCs.'*
This is asking you to show that you know *how* and *why* something occurs. You will need to *explain* enough to show that you understand what is going on, but most of your answer will be the reasons for and causes of whatever you are being asked about. In the example question, start by saying briefly what rural–urban migration is, where it occurs in LEDCs, and some of the consequences. Then take each cause in turn and show how the process works (e.g. push factors such as rural population growth, pull factors such as employment), then discuss each consequence (e.g. growth of shanty towns) and the reasons for it.

Discuss . . .

If you are asked to discuss a topic you will need to show a wide range of knowledge and understanding. It includes both *describing* and *explaining* the topic under discussion, but also drawing out some conclusions, e.g. do the advantages outweigh the disadvantages (or vice versa), what are likely to be future impacts, changes or developments? What could or should be done to deal with the issues raised?

Action point

Ensure that the case study you choose to answer a question is relevant to the topic – and only include those aspects of the case study that help answer the question. For example, if a question asks about social issues arising from nuclear power, don't write too much about the environmental impacts of your case study of Three Mile Island.

Compare/Contrast . . .

e.g. *'Compare and contrast the flood hydrographs of an urban and a rural catchment.'*
This command is asking you to look at two (or more) different situations and show how they are both similar and different. You will need to describe the similarities and differences but also be able to explain them, and show the consequences of any significant differences. In this example question you would describe differences in the hydrographs (e.g. shape,

peak flows, lag times, flooding risk) and then explain them (e.g. surface permeability, vegetation cover, etc.).

Critically evaluate...

e.g. *'Critically evaluate the range of solutions available for reducing soil erosion caused by agricultural practices.'*

This is asking you both to describe and to explain a topic and also to draw some conclusions on the basis of evidence. You will need, for this example, to discuss each solution available and indicate its pros and cons, and decide how likely it is to be a success, presenting evidence for your decision. Remember that 'being critical' involves looking at both positive *and* negative things, not just the negative ones.

Other key words

Using the right examples

Check the question to be sure that you are using the right examples. Is the question asking for a case study or examples? Does it want one or two? Should those examples be drawn from particular locations, e.g. 'with reference to a city in an LEDC country . . .' is very clear about what it is asking - write about Nairobi, or São Paulo, but not about London or Tokyo.

Using the right key concepts or ideas

Many questions are asking you to show that you understand particular concepts, and you need to be sure that these are the ones you cover in enough detail, e.g. 'with reference to a range of urban structure models . . .' is asking you to look at several models you know and could critically review (e.g. sector models, LEDC urban structure models, etc.) – do not write too much about urban growth except about how it helps explain urban structure.

Using the right geographical skills

Questions ask you to draw on particular skills, either by asking you to draw a sketch map or a diagram or by drawing on examples from your fieldwork. Be sure to include this if it is asked for, as there will be marks awarded specifically for this.

Action point

In addition to underlining command words, underline topic words in a different colour, i.e. words that define the topic you must address (e.g. with reference to population growth in LEDCs, discuss...)

Action point

As part of revision, practise drawing 'two-minute sketch maps' for each of the case studies you are learning, i.e. well-annotated sketch maps that can be drawn in two minutes.

Action point

Look through questions from sample papers. Then (a) identify suitable case studies to use for each, and (b) draw a spider diagram to show what ideas you would include in your case study. For each, check back to the question to be *sure* it is relevant.

Examiner's secrets

An excellent answer that shows a lot of understanding and knowledge will still score low marks if it doesn't answer the question that was asked.

Effective revision notes

The notes that you have taken throughout your study programme are the key to your preparation for the examination. They not only provide the information and material that you should learn but also provide a way of organising your knowledge to make it easier to understand and remember.

What should be in revision notes?

By the end of a course unit your notes should contain the following:

→ ideas (concepts) and information (knowledge/content) from your work in class and private study
→ details of case studies and examples to illustrate the ideas and principles of the topic studied
→ your answers to any test questions or exercises you have done.

Action point

How could you abbreviate the following in your notes:
→ 'because'
→ 'the government'
→ 'increases'?

Making notes

Remember that revision notes are notes, not full, grammatically correct essays. They need to be fully understandable by you, but not necessarily by anyone else. They should:

→ use abbreviations and your own shorthand way of writing things, e.g. write 'pptn' rather than precipitation, 'ag' instead of agriculture, or use text message language if it helps.
→ use symbols/signs to show ideas, e.g. use → instead of 'causes'
→ be organised in an easy way to read, e.g. use bullet points or lists
→ be visual, e.g. use spider diagrams or concept maps to show how ideas are linked, or use annotated maps or diagrams to summarise ideas. The figure below shows such a concept map.

Action point

Watch a TV documentary with a friend. Both take notes, then compare notes to see how each of you can learn from the other's style. You could try this with your whole class.

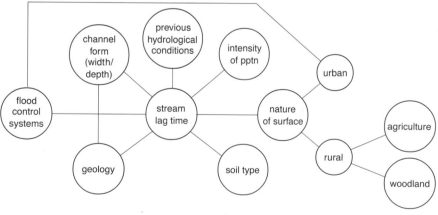

Making notes from lectures (including video/DVD)

Action point

List the geography units in your course. Against each one list the case studies you have looked at during your course.

→ do *not* attempt to write down every word as if by dictation
→ write down key words or facts as they occur
→ make sure you don't miss the next idea by being too busy writing
→ at the end of the lecture/lesson, take 2 or 3 minutes to go back through the notes and underline or highlight the main ideas (*not* the main facts).

Making notes from written resources

The secret of effective notes from reading resources is in the way you do the reading.

→ Never start taking notes until you've finished reading the resource.
→ Start by reading the introductory paragraph and the concluding paragraph, to give you a sense of what the item is about (and whether it is of any use to you!).
→ Then read the whole item carefully.
→ Summarise the resource in the following way:
   → Note the title, author, date and source of the resource.
   → Write a one-sentence summary of the resource.
   → Summarise the main ideas, using bullets, spider diagram, annotated sketch map or whatever is best for you.
   → For each idea, write down one specific example.
   → Keep your notes to no more than 20% of the length of the original, i.e. summarise a five-page article in no more than one side of notes.

Revising from notes

Remember that the aim of revision is not to clog your mind with facts but to ensure that you understand the concepts and ideas you have covered, with enough detailed knowledge to illustrate and exemplify those ideas. In revising from your notes:

→ start by reading through all your notes on a topic or unit to re-familiarise yourself with what you have covered
→ identify the key ideas you should understand in the topic by reading the relevant section of the specification that you are studying
→ for each key idea in the specification, produce a spider diagram or list of bullet points of the relevant things in your notes, e.g. for LEDC urbanisation you may have notes from two lessons, notes from a dvd on São Paulo, notes on São Paulo from a *Times* article, notes on a *Geography Review* article on Nairobi and Cape Town
→ then revise each key idea by reading through the relevant sections of your notes – you may need to do this 5–10 times
→ after each reading of your notes, try to recall as many of the key points as possible by writing the ideas down on paper
→ to revise the skills recorded in your notes, try practising each one, e.g. practise a chi-squared test, using some data in a textbook
→ look at past papers and questions to identify whether your understanding from your notes would enable you to answer them – if not, go back to a textbook to take some more notes.

For the final stages of revision, you may find it useful to summarise the main points and details of case studies for each key idea from the specification onto a single 'card index' card.

Action point

Organise bullet point lists so that the first letters spell a word or are in alphabetical order. This makes memory easier, e.g. river erosion occurs through CASH:

→ **C**orrasion
→ **A**ttrition
→ **S**olution
→ **H**ydraulic action.

Action point

Draw a spider diagram to show the causes and consequences of global warming.

Examiner's secrets

Some case studies link to several topics in your course, e.g. agriculture in Kenya may relate to agriculture, development, environmental hazards, tropical savannas, etc. Add a note to your notes in every relevant place to look at such case studies.

"Knowledge should be a light to the mind not a burden on the memory."

Anon

Examiner's secrets

Notes are good for revision, but your written answers must be in full sentences and correct English. Only use notes as annotations to diagrams.

Index